Biological Efficiency
in Agriculture

Biological Efficiency
in Agriculture

C. R. W. SPEDDING, J. M. WALSINGHAM
and A. M. HOXEY

Department of Agriculture and Horticulture,
University of Reading, England

1981

ACADEMIC PRESS

A subsidiary of Harcourt Brace Jovanovich, Publishers

London   New York   Toronto   Sydney   San Francisco

ACADEMIC PRESS INC. (LONDON) LTD.
24/28 Oval Road,
London NW1

United States Edition published by
ACADEMIC PRESS INC.
111 Fifth Avenue
New York, New York 10003

British Library Cataloguing in Publication Data
Spedding, C. R. W.
  Biological efficiency in agriculture.
  1. Field crops — Physiology
  I. Title II. Walsingham, J. M. III. Hoxey, A. M.
  631.5 SB185.5 80–42231
  ISBN 0-12-656560-0

Filmset by Oxford Publishing Services and printed in Great Britain by John Wright & Sons
(Printing) Ltd. at The Stonebridge Press, Bristol

To all those
who collected the data

PREFACE

This book has two main purposes. The first is to examine, systematically, the useful meanings of biological efficiency in an agricultural context and to illustrate the uses of the more important efficiency ratios.

Agriculture is nearly always concerned with efficiency, and production has to be expressed per unit of one or more resources, but it is not a simple matter to decide which of the many possible ratios are most appropriate for any particular purpose and it is not always possible to obtain the data needed to calculate some of the ratios of interest.

Because of the significance of biological efficiency in agriculture, it is also an important way of looking at the subject. This, then, is the second purpose of the book, to present a comprehensive picture of world agriculture in terms of the efficiency with which the major resources are used to produce the major products. Such a picture is one of the main ways in which world agriculture can be described, provided that the efficiency concept is related to agricultural processes or to biological processes in an agricultural context. The book does not deal in detail with those biological processes, such as photosynthesis, that cannot be described as essentially agricultural, however important they may be *within* agriculture.

Biological efficiency has been used here rather in the sense of the technical efficiency of biological processes that are involved in agricultural production. Such efficiency has economic significance even when it is not expressed in monetary terms: indeed, it can be argued that financial expressions would inevitably be of only transitory value.

Since our aim has been to provide a picture of world agriculture from one important point of view, we hope that the book will be of interest to a wide range of readers, but especially to students of agriculture, geography and those biological subjects of relevance to agriculture.

Department of Agriculture
and Horticulture,
University of Reading,
Earley Gate,
Reading

C. R. W. Spedding
J. M. Walsingham
A. M. Hoxey

January 1981

ACKNOWLEDGEMENTS

It is a pleasure to acknowledge the help we have received during the preparation of this book from Dr N. R. Brockington.

We are grateful for permission to use published material from the following sources:

Professor E. H. Roberts and the Administrator UK International Solar Energy Society (Table 3.2); J. M. Paturau and Elsevier Amsterdam (Fig. 12.1); Dr J. L. Corbett and Pergamon Press Limited (Fig. 17.3); R. A. Terry and The British Grassland Society (Fig. 17.4); A. S. King and Academic Press (Fig. 17.5); Professor J. K. Loosli and McGraw-Hill Book Company (Figs 17.7 and 17.8); Dr C. C. Balch and Elsevier/North Holland (Fig. 18.1); Oxford University Press (Figs. 18.3, 19.1 and 23.1); Academic Press (Fig. 18.5); R. V. Large and Plenum Press (Figs 19.2, 20.4, 20.5, 20.6 and 20.7); Professor J. C. Bowman and Cambridge University Press (Table 20.7); The Chemical Society (Table 20.10); Dr W. O. McLarney (Table 21.7(a)); H. Ackefors, C. G. Rosen and AMBIO (Table 21.7(b)); L. Norman and Longman's (Figs 22.3 and 23.4); R. V. Large and Cambridge University Press (Fig. 23.2); Longman's (Table 23.16); A. D. James and Longman's (Table 23.17); Dr Eva Crane, International Bee Research Association (Tables 24.3 and 24.4); Elsevier, Amsterdam (Table 28.1). We wish to thank Mr W. R. Buckley for his help in the construction of Fig. 11.1.

Our thanks are also due to Mrs Valerie Craig for typing the several drafts and helping in many ways during the course of preparation.

C.R.W.S.
J.M.W.
A.M.H.

CONTENTS

Part III

Efficiency in Animal Production

Part IV

The Relative Efficiency
of Plants and Animals

Part I
Introduction

1

BIOLOGICAL EFFICIENCY

Efficiency

There is general agreement that efficiency can most usefully be defined as the ratio of output to input. Biologists have tended to denote efficiency with the letter "E" (Spedding, 1973), although physicists and engineers, who have employed the concept for much longer, have adopted the Greek "η", reserving "e" for energy. We shall use "E" in this book and express the definition as $E = O/I, O$ representing a chosen output and I a chosen input.

Such a ratio can apply to innumerable combinations of output and input and each of these can be expressed in many different terms, some of them allowing several different outputs (or inputs) to be combined. The fact that efficiency can mean many different particular things does not reduce its usefulness: it is no different, in this sense, from words like "animal", "plant", "aeroplane" or "book". The general concept is clear and obviously of value, and part of its value lies in its applicability to a great many, different, particular cases.

Particular versions are often named, such as energetic efficiency, relating energy output to energy input, and it is sometimes thought that outputs and inputs should be expressed in the same terms. This is not necessary, however, and unnecessarily limits the usefulness of the concept. Furthermore, everyday usage of the term already includes ratios in which output and input are of a quite different kind, quite apart from differences in the terms employed. Familiar examples are "miles per gallon", "words per minute", "miles per hour", "gallons per cow" and "protein per acre".

There is no question of efficiency being good or bad, therefore, or of greater efficiency being beneficial or disadvantageous, nor of one ratio being right or another wrong, except for specified purposes. If we are interested in milk production, then miles per gallon and words per minute are clearly irrelevant, but several different ratios may be relevant and these can be expressed in several different terms. Once our interest has been rigorously specified, however, the most relevant ratio will be fairly obvious.

3

Specification of interest has to flow from a definition of purpose, why we are interested and what for, how the ratio is to be used and what we think it is going to tell us. Confusion often occurs here, simply because there are usually several purposes involved and one may not clearly over-ride all others. So it may be necessary to choose the most useful of the various possible and relevant expressions of efficiency, or to accept that several have to be considered at the same time. Very often, judgement has then to be exercised in order to weight each ratio and also to take account of other factors, constraints and limitations.

Biological Efficiency

There is no essential difference in applying the concept of efficiency to biology and to other applications. The simplest view is that biological efficiency represents the efficiency of a biological process and, since the latter could use physical or chemical inputs, a biological efficiency ratio could involve non-biological inputs. Similarly, the results of biological processes may also be non-biological (e.g. a pearl, a territory, a house-martin's nest, a foot-print, a desert): this notion has to be even further expanded if Man's activities are included.

However, biological efficiency has also been used to describe the biological success of organisms and this generally has to be expressed in biological units. For example, the success of a species is usually described in terms of its numbers, of its rate of population increase, or the number of different habitats or niches it can occupy, or its competitive success relative to other species. These are, of course, legitimate uses but they are not excluded by the wider view and there are a great many ratios of interest that could not be included if *biological* success was made an essential part of biological efficiency.

The wider view, then, would simply define biological efficiency as the efficiency of a biological process or processes. This means that we can assess the efficiencies of combined processes, even including such complicated combinations as ecosystems. The latter include agricultural systems and thus allow the determination of biological efficiency within Agriculture. There is nothing fundamentally different in the calculation of the efficiency of a biological system and that of one that is not biological. In many cases, the processes may be very similar. It is also worth remembering that biological systems commonly use non-biological inputs, such as solar radiation, and that some of the most relevant efficiency ratios will be based on them. Indeed, as soon as one considers solar radiation, it is clear that efficiency is already an established concept in biology. It is certainly true that Agriculture is greatly concerned with efficiency and, since it is based primarily on the harnessing of biological processes, the efficiency of Agriculture is greatly dependent on biological efficiency — though not wholly so.

A Definition of Agriculture

Agriculture is rarely defined, partly because most people already consider that they know what it is and partly because of the difficulty in arriving at a useful definition. There is a sense, of course, in which definitions tell you what something is not, i.e. they enable you to distinguish something from other things, but they do not, comprehensively, tell you what it is.

The following definition serves to distinguish Agriculture from other things: it defines it as an activity and distinguishes it from other activities.

Agriculture is an activity of Man, primarily aimed at the production of food, fibre and other materials (as well as power and fuel), by the controlled use of (mainly terrestrial) plants and animals.

This definition includes other recognizable activities, such as forestry, horticulture and fish-farming, but does not mention the use of soil or sunlight, both normally characteristic of Agriculture. Nevertheless, the definition does identify the essential attributes of Agriculture, namely Man's use of biological processes (involving plants and animals), which characteristically involve both soil and sunlight.

No such activity could continue if it was grossly inefficient in the use of costly resources. If the value of its purchased inputs was greater than the value of its usable outputs, it could not be sustained for long. Agriculture is always concerned, therefore, about the efficiency with which such inputs are used. But *all* processes are energetically inefficient if all sources of energy are counted. Energy is always being used and dissipated as heat during the biological processes themselves.

Agriculture thus requires a source of energy costing much less than the energy value of its products, if it is to make sense in energetic terms. Of course, it also produces many other things as well as energy (see Chapter 2). Even so, the total energy inputs to Agriculture are very large and a very cheap source of energy is imperative.

This is provided by solar radiation and agricultural systems are often described in terms of the conversion and use of solar radiation to produce crops and animal products. Figure 1.1 shows an example of this kind of energy flow in agriculture. This does not include any human activity, however, and could be a non-agricultural system. Figure 1.2 does include Man and also includes examples of the use of "support" energy. The latter embraces the "fossil" fuels (coal, gas, oil, petrol), originally derived from solar radiation, which have been used increasingly in developed agricultural systems, directly as fuel and, indirectly, in the manufacture of inputs and the processing and distribution of products (see Chapter 2).

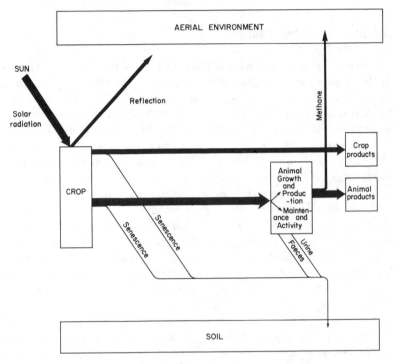

FIG. 1.1   An example of energy flow in agriculture

One of the reasons for these inputs of support energy has been to increase the extent of Man's control over the biological processes in Agriculture. A very simple example is a wire fence: this increases the degree of control of farm animals but costs support energy to manufacture and, possibly, to erect. Very often the form of control has been over nutrient supply, in order to remove restrictions on plant growth. The manufacture and application of artificial fertilizers are an example of this. One result of these extra inputs of energy is to increase the efficiency with which land and, thus, solar radiation are used, by greatly increasing yields per unit area. These aspects of efficiency will be discussed in the relevant chapters, dealing with each major group of products in turn. Here, we simply wish to draw attention to the way in which this increasing degree of control has limited the use of biological processes in Agriculture, either by constraining them to a narrower role or by more or less complete substitution by non-biological processes (Table 1.1).

There is no implication here that any of these developments are right or wrong and it would clearly be ridiculous to generalize about them in this way. Some, such as the controlled slaughter of animals by Man and the elimination, so far as is possible, of predation by other carnivorous animals, are an essential part of the agricultural activity. Others, such as artificial incubation of hens' eggs instead of natural incubation by the hen, make it possible to increase productivity per hen and to reduce losses. In these cases, an economic calculation may be required to decide whether something is sensible or not and in what economic (and biological) circumstances this is so.

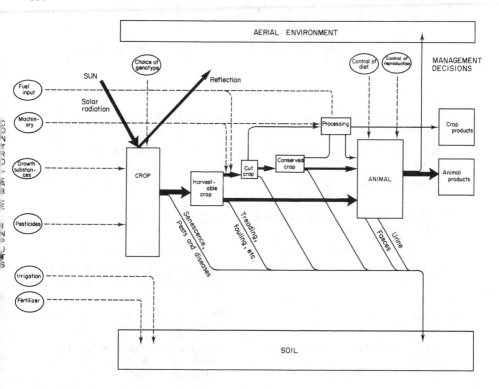

FIG. 1.2   An expanded version of Fig. 1.1 to include human activity

There are other examples, however, such as cattle feed-lots, where the issues are more difficult to resolve. Here, all feed is brought to the animal, which, naturally, could collect its own, and all excreta is removed, when naturally the animal would arrange distribution by itself. When support energy is cheap and plentiful, the additional costs may be sensibly balanced by a reduction in harvesting inefficiencies and supervisory costs. It may still be argued that it makes no sense to have produced, in this way, a costly dung-disposal problem with water pollution risks: on the other hand, the collection of faeces at one place may make it possible to recycle the material in a more efficient way than would occur in natural grazing. Feed-lots are an interesting example of the way in which economies of scale for some processes may be accompanied by diseconomies of scale for others. This clash of efficiencies means that there are optimum values for size and structure of enterprises (Wilson, 1976).

Some will argue that there are equally clear cases where the substitution processes have gone too far, either on the grounds of animal welfare (that the animal's natural functions have been too far curtailed), or on the

TABLE 1.1   The displacement of (a) biological processes and (b) of man- and animal-power, in agriculture

(a) Biological process displaced	Non-biological process substituted
Natural fertilization in plants/seed dispersal	Plant breeding/harvesting of seeds
Fixation of atmospheric N by bacteria	Application of artificial nitrogenous fertilizers
Exploration of soil by roots for potash and phosphorus and water	Application of artificial fertilizers; irrigation
Natural control of pests and weeds	Use of pesticides and herbicides
Collection of feed by animals	Harvesting, processing and automated provision of compounded feed; forage conservation
Grazing	Zero-grazing (the cutting and carting of herbage)
Natural deposition of excreta on the land	Collection of excreta from housed animals and its disposal, treatment or distribution on land
Incubation of eggs by hen birds	Artificial incubator
Natural service by male animal	Artificial insemination
Natural hormonal processes	Control of light, day-length and temperature; use of synthetic hormones
Natural suckling (of calves and lambs)	Artificial rearing on milk substitutes
Natural immunity to disease in animals	Use of vaccines
Use of animal power	Use of machines and fossil fuel

TABLE 1.1 (*cont.*)

(b) Operation displaced (based on man- or animal-power)	Processes or equipment substituted
Herding and guarding of animals	Fencing and housing
Feeding of animals	Machinery
Watering of animals	Automatic water supply
Hand milking	Milking machines
Carting and spreading of manure	Tractors and machinery
Hand, horse, oxen or buffalo cultivation sowing planting ploughing weeding	Tractors and machinery Direct drilling and herbicides Herbicides applied by machinery
Harvesting of crops	Tractors and machinery
Hand application of fertilizer	Machinery and tractors
Drainage ditch digging	Machinery

grounds that substitution always involves more support energy and that this is costly, in short supply and, in any case, non-renewable.

It is certainly the case that improvement in the efficiency of use of support energy probably requires an increased emphasis on biological processes, whilst, at present, increases in the efficiency of use of solar radiation require increasing inputs of support energy. This generalization will not apply to all agricultural systems in all parts of the world, but will be true of the majority.

It is most important, however, not to oversimplify either the calculation or the interpretation of efficiency ratios.

The Calculation of Efficiency

It is obvious that efficiency cannot be calculated without specifying both the inputs and outputs of interest. There are three other important features that must be specified in addition. They are (a) the process or system whose efficiency is being assessed; (b) the context or environment in which it is assumed to be operating; and (c) the period of time to which the calculation refers.

(a) Where a simple biological process is being considered, there may seem to be little problem in specifying it. It is not usually a simple matter to be so specific, however, as the following example shows.

Consider a plant growing in soil and producing protein in its leaves. Nitrogenous fertilizer may be added to the soil and the efficiency of nitrogen output calculated per unit of nitrogen input. But the soil may contain nitrogen already and the additional quantities may be large or small relative to the initial amount. Or the plant may be a legume or be associated with other plants able to fix atmospheric nitrogen. Clearly, the meaning of the efficiency ratio is going to differ according to exactly where the boundaries are drawn around the system considered.

Another example illustrates further the problem of the system content of the particular input selected. Suppose we wish to calculate the efficiency with which feed energy is converted to milk energy by cows. Clearly we have to specify the kind of cow (its breed, age, weight, stage of lactation) and the kind of feed (digestibility, energy content, toxicity, dry matter content), but the fat reserves of the cow may be difficult to assess. Yet a significant proportion of the milk energy may be derived from these reserves and not from the feed, at least for a time, in a high-yielding or a poorly-fed cow.

(b) No system or process operates in a vacuum and it is unlikely that the entire context will be included in the system specification. Most descriptions of agricultural systems and processes, for example, will probably not mention oxygen, not because it is not an essential component but because it is assumed that all relevant environments will include it in non-limiting concentrations.

Similarly, features of the environment, such as temperature, humidity and topography, may not be specifically mentioned but will be assumed to be suitable. Thus, the calculation of efficiency of feed conversion by fish may not actually include any reference to water. Yet it is obvious that the efficiencies of cacti and pondweed will depend on the environmental context, as will the efficiencies of sheep, goats, cows, reindeer and pigs. No efficiency ratio can have much meaning if the environmental context is not specified.

(c) Efficiencies can legitimately be calculated over any period of time and comparisons are not always best made over the same absolute period. It does not really make sense to compare the reproductive efficiencies of mice and elephants over exactly the same period but, if this is to be done, the period chosen must be long enough for the elephant. Very often, lifetime performances may be appropriate and these will vary with the species.

Where land use is being considered, it is often sensible to take into account natural periodicities of growth: seasonal and annual cycles can be used. With animals, breeding cycles may mark off periods of time that are comparable for different species.

Whatever period of time is chosen, it is clearly necessary to specify it, since interpretation depends upon this.

The Interpretation of Biological Efficiency

Given a sufficiently complete specified calculation, interpretation should be straightforward: the results of the calculation should serve the purpose for which it was made.

Purposes will tend to be of two main types. The first is simply to decide which process or system is more (or most) efficient at doing a specified thing in a specific context. The object is to make it possible to choose the most efficient of those considered. Frequently, however, it will be desirable to know what effect on efficiency will follow changes in some of the most important attributes of the system or the environment.

The second purpose is to assess efficiency in such a way that it can be improved if it is not satisfactory. A simple ratio does not allow this.

For these further purposes it is necessary to expand the ratio, to show what factors are important determinants of it and to determine the consequences of varying them over relevant ranges of values.

Figure 1.3 illustrates how an efficiency ratio may be expanded and indicates how variables may interact with each other (for a systematic treatment of this, see Spedding, 1975).

There are two main, important conclusions to be drawn from these considerations.

The first is that a single figure for the efficiency of a process or a system, can only apply to a highly specific situation. The figure will be different for changes in a whole range of variables and efficiency can be expressed as a series of response curves to the most relevant changes in factors that are likely to change or can be controlled.

The second is that, if efficiency calculations are to be used as a basis for change and improvement, the process or system has to be described in sufficient detail and in such a way that its sensitivity to change in important variables can be adequately expressed. Furthermore, this has to be in sufficiently quantitative terms. To do this requires some form of modelling, ultimately in mathematical terms.

The advantages of modelling in the study of agricultural systems have been discussed at considerable length recently (Dalton, 1975; Spedding and Brockington, 1976) and they apply here just as much as anywhere else.

Indeed, if Agriculture is nearly always concerned with efficiency, since we are rarely interested in any output except in terms of how much is produced per unit of some input, it follows that calculating efficiency is simply one way of describing the study of agricultural processes and systems.

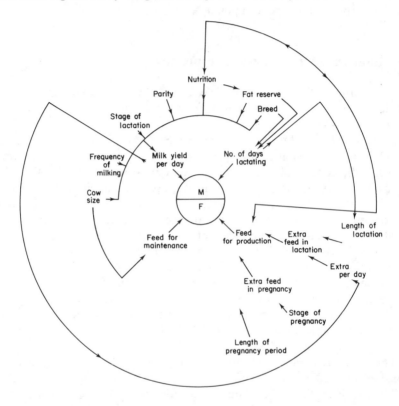

FIG. 1.3   The expansion of an efficiency ratio, illustrated by the factors affecting milk output (M) per unit of feed (F), for an individual cow over a period of one year

As any one of these processes and systems can be considered in so many different ways, in different environments and for different periods of time, the calculation of all possible efficiency values for each system is quite impossible.

Most of the subsequent chapters are therefore concerned with a discussion of the special or even unique features of biological efficiency in relation to the main groups of plant and animal products. Before proceeding to this sequence, however, it is worth considering, in some detail, the purposes for which Agriculture is carried out.

References

Dalton, G. E. (1975). "The Study of Agricultural Systems", Applied Science Publishers, London.

Spedding, C. R. W. (1973). The meaning of biological efficiency. *In* "The Biological Efficiency of Protein Production", (Ed. J. G. W. Jones), Cambridge University Press, London.

Spedding, C. R. W. (1975). "The Biology of Agricultural Systems", Academic Press, London and New York.

Spedding, C. R. W. and Brockington, N. R. (1976). Experimentation in agricultural systems. *Agric. Syst.* **1**, 47–56.

Wilson, P. N. (1976). The optimum size and structure of enterprise. *In* "Meat Animals" (Eds D. Lister, D. N. Rhodes, V. R. Fowler and M. F. Fuller), Plenum Press, New York.

2

AGRICULTURAL PURPOSES, PRODUCTS AND RESOURCES

The definition of Agriculture, given in Chapter 1, implied that the purposes for which it is carried out can be summarized as the production of food, fibre, power and fuel (and some other materials). This is probably true of any particular producer but may not adequately express the role of agriculture in world, or even national terms. It can be thought of as including the production of ornamental crops and even amenity horticulture but it leaves out of account, for example, the cultivation of plants for erosion control.

The Role of Agriculture

The main role of agriculture must be the conversion of raw materials into useful products, largely by the use of solar radiation. As already mentioned, however, the additional use of support energy is deliberate and purposeful, as would be the use of other sources of energy, such as wind and water power or nuclear fission or fusion. The purposes of agriculture can be described in terms of all its outputs, including products, and all its inputs (see Fig. 2.1).

Although it may be argued that agriculture involves production, this is clearly not always its purpose, never mind being its *main* purpose.

Many countries, including those of the E.E.C., may have substantial and embarrassing surpluses of particular products, yet the price is not necessarily allowed to fall in such a way that production is discouraged. There are many reasons for this, including the idea that such surpluses may be temporary and that the capacity to produce must be maintained: this would be a view that might well be held by a nation but is less likely to concern the individual consumer. The farmer would certainly see his own interest as best served by continuing to receive adequate prices for his products even during a period of over-production, in order that he should be able to remain in business.

Incidentally, any other economic framework is likely to act as a brake on

14

production, to prevent prices falling due to over-production, and will generally be aimed at a degree of scarcity wherever producers are able to organize this.

In developed agricultural systems, productivity per man is high and pride is taken in how few men are required per unit of output. In many under-developed countries, by contrast, a major purpose of agriculture is to provide employment. Very often, the purposes and interests of farmers, consumers and the nation will differ in quite important respects, and these issues become an important political matter to be resolved by Government action.

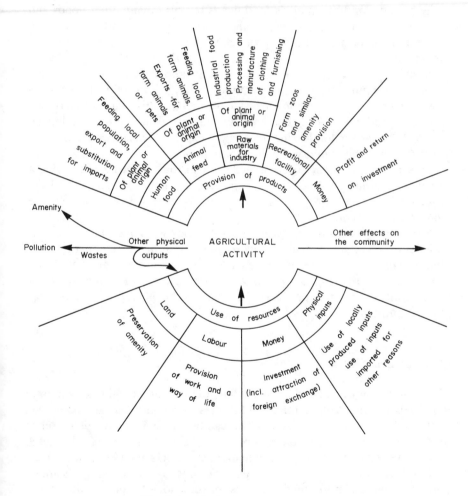

FIG. 2.1   The purposes of agriculture

There are many cases, then, when the use of resources may be a more important feature of agricultural purpose than the production of products.

In general, some aspect of production is likely to matter but it is not always the production of food as such. A subsistence farmer, living on his own produce, will obviously be concerned with how much food he produces (and at what cost). If he is selling surplus produce, however, he will be more concerned with its monetary value at the point where he can sell it. So it is quite possible for a farmer to produce less food for people, and even none for himself, but to use all his resources to produce high value products (not necessarily food at all) for sale to others. He can then afford to purchase not only his own and his family's food needs but also many other goods and services as well. His purpose is then monetary gain and this is quite commonly associated with the production of *less* nutrients for human consumption, per unit of land and other resources. This is well illustrated by the concentration on meat production rather than crops in countries where meat commands a much higher price per unit of energy or protein, although it is well known that more people can be fed per unit of cultivated land if it is used for crop production. The economic adjustments of such situations depend upon the *need* for food being adequately expressed as a *demand* but this cannot occur when hungry people are also poor.

It is not always easy to disentangle the various purposes underlying a particular agricultural activity and the reasons that have determined the choice of product, of resources and of the particular ways of producing the one from the other.

Since many reasons for considering efficiency will have to do with how efficiently a system is achieving its purpose(s), an understanding of what these may be is of some consequence. It also has to be remembered that agricultural activities may generate outputs that are incidental (or even acciuental) to the main products: these have to be included in a consideration of outputs.

Agricultural Outputs

These may be grouped under the headings of human food, animal feed, raw materials for industry, recreational facilities, waste products and pollution (see Tables 2.1 to 2.6), most of which can be conveniently divided on the basis of their plant or animal origins. Money is not mentioned (a) because any of the above can be expressed in monetary terms and (b) because once stated it cannot be further sub-divided. This may be an oversimplification and there may be important differences between, for example, profit, return on investment and capital appreciation, that should be distinguished.

It would certainly be an oversimplification to argue that agriculture does not directly produce money. Strictly speaking, it does not produce much food either: useful, palatable and digestible food is usually the end-point of a substantial chain of processing, preparation and cooking, with preservation and storage often intervening. To the extent that agriculture really produces raw materials for the food industry, it may be considered that whole food-producing systems should be examined when efficiency is being assessed.

TABLE 2.1 Agricultural outputs: human food from plants and animals

Crop products	Examples
Cereals	Wheat, rice, maize
Starchy roots	Potatoes, sweet potatoes, yams, cassava
Sugar	Cane, beet, palm, maple
Pulses	Beans, peas
Nuts	Groundnuts, coconuts, walnut
Oils	Olive, coconut, sesame seed, sunflower seed
Vegetables	
roots, bulbs, tubers	Beets, carrots, onions, potatoes
leaves	Cabbage, celery, spinach, lettuce
other	Cauliflower, beans, peas, tomatoes
Fruits	
starchy	Bananas, plantains
citrus	Lemon, orange, grapefruit
fat-rich	Avocado, olive
other	Apples, grapes, mangoes, melons, plums
dried	Dates, figs, prunes, raisins
Beverages	
non-alcoholic	Tea, coffee, cocoa, maté, soft drinks, cola
alcoholic (beer, spirits, wine)	Barley, hops, potatoes, grapes

TABLE 2.1 (*cont.*)

Flavourings	
spices	Peppers, mustard, cloves, vanilla, nutmeg
herbs	Parsley, rosemary, mint, thyme
tuberous roots	Ginger, horse-radish, liquorice, turmeric

Animal products	Examples
Meat	Beef, lamb, pork, camel, buffalo, goat, horse, rabbit
Poultry	Chickens, turkeys, geese, ducks
Game birds	Pheasant, partridge, grouse, Guinea fowl
Offal	Liver, heart, kidney, brain
Blood	Cattle blood, sausages
Eggs	Hen, duck, goose, turtle
Fish	Trout, carp
Shellfish	Oysters, mussels, shrimps, prawns
Other aquatic products	Frogs, turtles
Milk and milk products	Whole milk (cow, goat, sheep, buffalo, camel), cheese, yoghurt
Fats	Butter, ghee, lard, suet, tallow
Sugar	Honey
Miscellaneous	Snails

Sources: Masefield *et al.* (1969); Rao (1976)

There is much to support this view: certainly it is one of the many views that may be necessary to build up any kind of complete picture of agriculture in the service of man. From the farmer's point of view, however, the natural end-point for his production systems is when the product is sold.

TABLE 2.2 Agricultural outputs: animal feed[a]

From plants	
Feed	Examples

Fresh green feed	
Grasses	Ryegrasses, guinea, pangola, wheatgrass
Forage legumes	Clovers, lucerne, stylo
Cereals and grain legumes used green	Oats, maize, peas, beans
Brassicas and other leafy crops	Cabbage, kale, rape
Leaves of roots and waste from vegetable food crops	Sugar beet tops, peas and bean haulm
Leaves and shoots of trees and shrubs	Acacia (Mulga, boree), saltbush, heather
Bulky conserved feeds	
Hay	Grass, legume, cereal
Dried crops	Grass, legume
Silage	Grass, legume, cereal, crop by-products
Roots	Potatoes, swedes, turnips
Bulky by-products	Sugar cane bagasse, beet pulp, straw, chaff, bran, citrus pulp and skins, olive residue
Concentrate feeds	
Cereal	Barley, maize, millet
Pulses	Beans, peas, lupins
Oilseeds, cakes and meals	Linseed, soyabean, groundnut meal
Miscellaneous nuts and seed	Acorns, chestnuts, locust beans

From animals	
Feed	Examples

Meat by-products	Meat, bone, blood and feather meals Offal Tallow and other fats
Milk and milk by-products	Whole milk ⎫ Skim milk ⎬ liquid or dried Butter milk ⎭ Whey
Faeces	Processed or fresh poultry, cattle or pig faeces
Worms and other invertebrates	

[a]Feed for agricultural animals and other domesticated animals (e.g. dogs, cats, horses, donkeys, birds, fish)

TABLE 2.3 Agricultural outputs: products for human use other than food

Human use industry	Source of raw materials	
	Plants	Animals
Construction	Timber Thatch	Faeces (for bricks) or as plastering
Furniture	Timber, kapok	Hair, leather
Pulp and paper	Wood, straw, kenaf	
Fuel	Wood, charcoal, straw, plant fuelcrops, kenaf	Faeces
Textiles and clothing	Cotton, flax, jute, hemp, ramie, sisal, coir	Wool, hair, furs, skin, silk
Household goods	Wooden and woven utensils, cork, rubber	Leather, bristle, bone, glue, tallow, beeswax
Footwear	Cork, rubber, wood	Leather, skins, fur
Industrial oils	Castor bean, tung, linseed, coconut, oil palm	Fats
Tobacco	Tobacco	
Pharmaceuticals	Opium, quinine	Hormones, heparin
Fertiliser	Stubbles, green manures, compost	Blood, bones, horn
Dyes and tannins	Indigo, bark, wood	Gall bladder
Insecticides	Roterone, pyrethrum	
Perfumes and cosmetics	Rose oil, jasmine, violet, patchouli, orange blossom, sandalwood	

TABLE 2.4   Agricultural outputs: recreation

Recreation	Agriculture's contribution
Walking	Footpaths, landscapes
Riding	Breeding and rearing horses, production of feed, space to ride, e.g. bridle paths
Holidays	Accommodation, e.g. bed and breakfast. Space and facilities for campers and caravaners.
Fishing	Letting of fishing rights for private rivers and ponds. Management of these waters to provide good fishing.
Shooting	Letting of shooting rights on private land or water. Management to encourage wild animals and birds, e.g. pheasants, partridge, grouse, wildfowl, rabbits, deer, etc. and deliberate stocking with large numbers, e.g. pheasants.
Education	Farm parks and zoos
Provision of food	Self-pick food
Pleasure	Visual amenity, hedges, flowers, trees and pleasing smells

TABLE 2.5   Agricultural outputs: wastes

Animal wastes

Excreta from	Cattle Pigs Poultry
Slaughtering waste	Intestines, etc.

Plant wastes

Non-harvested parts	Grain losses by shedding. Roots or tubers left in ground. Parts of plants not harvested — maize stover, sprout and tomato plants, etc.
Rejected parts	Straw — wheat, rice, etc. Bean and pea haulm Root and tuber tops Vegetable processing waste Sugar cane tops and bagasse

TABLE 2.6 Agricultural outputs: pollution

Pollutant	Source
Chemicals in excess e.g. nitrates, phosphates or copper	Fertilizer, faeces and urine, herbicides, fungicides and insecticides
Noise	Heavy machinery; tractors, driers, harvesting equipment Animals at weaning or feeding time
Dust	Feed mixing, cultivations in dry conditions, combine harvesting, grass drying
Smoke	Straw-burning
Odour (unpleasant)	Excreta from housed animals under some storage conditions Silage
Organic matter entering waterways and causing deoxygenation	Untreated animal wastes or silage effluent

The Value of Products

Some of the outputs of agriculture are extremely difficult to value in any terms. An attractive view (in scenic terms), for example, differs in value to different people and at different times and what you are prepared to pay for something depends upon your wealth and a host of other factors.

Similarly, the negative value of something like pollution varies between people: indeed, not everyone would be agreed about whether something is a pollutant or not and, of course, anything can be a pollutant in given circumstances.

Products, by and large, can be valued and some are generally more valuable than others, although prices, both relative and absolute, may fluctuate quite widely.

An example of the effect of the nature of the product on its price is the difference between the price of 1 kg of protein from animal and plant sources (Table 2.7). This clearly reflects preferences, beliefs and traditions and could change in the future. Nevertheless, there are important distinctions to be made between the attributes that make a food nutritionally useful and those which make it attractive.

In other cases, price relativities exist but it is extraordinarily difficult to establish a sound basis of comparison. The relative prices of sheep meat and wool are an example of this. Meat and wool are both produced by sheep, with either product being the major one and the other a by-product, depending primarily on the breed of sheep and the harshness of the environment. So the proportion of a sheep farmer's total income that is derived from wool varies considerably from 6% in lowland U.K. to 100% in parts of Australia, for example (McKay, 1962; Kilkenny, 1976).

Even so, the price of wool relative to that of sheep meat does not vary to anything like this extent (Table 2.8).

TABLE 2.7   The price of 1 kg of protein (with whatever energy is present) from animal and plant sources in 1973

Product	World prices[a]	Wholesale prices	Retail prices[d]
Wheat	22p	40p[b]	
Wheat flour	37p		
Rice	94p		
Flour			98p
Bread			£1.64
Groundnuts		50p[b]	
Soyabeans	14p	19p[b]	
Peas		£1.38[b]	£2.74
Runner beans			£2.56
Potatoes	£1.17	£1.86[c]	£2.52
Cocoa	£1.26	£2.07[b]	
Beef	£3.22	£5.71[c]	£8.99
Poultry meat	£1.45		
Bacon		£5.27[b]	
Milk	£1.95 (evap)	£1.78[c] (whole)	£2.85
Cheese	£1.94		£2.79
Eggs		£2.76[c]	£8.70
Cod			£5.81
Herring			£2.78

[a]F.A.O. (1973) World average export unit values   [c]U.N., F.A.O. (1974)
[b]The Public Ledger (1973)   [d]Department of Employment (1973)

In such circumstances, it is clear that economic and physical efficiency ratios will often diverge greatly. The point is well illustrated by the discrepancy between the energy content of human labour and its unit cost.

TABLE 2.8 Ratio of wool production to sheep meat production, and wool price to sheep meat price in selected countries (1973) (from Kilkenny, 1976)

	kg wool produced per kg sheep meat produced	Ratio of sheep meat[a] price per kg to wool price per kg
Argentina	0.68	—
Australia	0.64	0.40
New Zealand	0.40	0.39
United Kingdom	0.14	1.13
France	0.09	0.76
West Germany	0.13	—
Irish Republic	0.21	0.89
Italy	0.20	—
Poland	0.61	0.25
Spain	0.09	0.65
Turkey	0.08	—
United States	0.15	0.96
U.S.S.R.	0.28	—

[a]Producer prices

Agricultural Resources

The most important resources are usually regarded as land, labour, money and physical inputs. Although the last term is fairly elastic, this list is really quite inadequate. It represents resources that have to be paid for, or can be controlled, and thus does not include atmospheric oxygen and nitrogen, and ignores the biological content of the systems that are used agriculturally. This may be partly due to distinctions between the *content* of the system, including its animals and plants, and resources used *by* the system.

Land is certainly an important resource and may be taken to include areas of water, whether farmed or not.

Labour is a somewhat misleading-sounding category: it needs to include managerial and other skills, as well as physical work, and in the latter context should take account of animal power.

Physical inputs obviously include machinery, buildings and fertilisers but energy (solar and support) has also to go in here if no special category is added for it. It would probably be better to list energy separately, however, and the same is true of the animals and plants, although it is difficult to distinguish between those which are essential to an agricultural system and those which are not.

The major agricultural animals and plants can be listed (see Tables 2.9–2.14) but the other organisms that are essential to many agricultural operations are equally resources of some importance. It would be impossible to list them but examples are given in Table 2.15.

TABLE 2.9 The major agricultural mammals

Cattle	*Bos taurus, Bos indicus*	Camels	*Camelus dromedarius*
			Camelus bactrianus
Buffalo	*Bubalus bubalis*	Alpaca	*Lama pacos*
Musk ox	*Ovibos moschatus*	Llama	*Lama glama*
Yak	*Bos (Poephagus) grunniens*		
Sheep	*Ovis aries*	Rabbits	*Oryctolagus cuniculus*
Goats	*Capra hircus*	Guinea-pigs[a]	*Cavia porcellus*
Red deer[a]	*Cervus elaphus*	Capybara	*Hydrochoerus hydrochoeris*
Eland[a]	*Taurotragus oryx*	Mink[a]	*Mustela vison*
Horses	*Equus caballus*	Pigs	*Sus scrofa*
Asses	*Equus asinus*		
Mules		Dogs	*Canis familiaris*

[a]Not numerically important

TABLE 2.10 Agricultural birds (Only the first four[a] can be considered as of major importance.)

Chicken[a]	*Gallus gallus*	Grey partridge	*Perdix perdix*
Turkey[a]	*Meleagris gallopavo*	Red-legged partridge	*Alectoris rufa*
Duck[a]	*Anas platyrhynchos*	Red grouse	*Lagopus lagopus scoticus*
Goose[a]	*Anser anser*	Ostrich	*Struthio* spp.
Guinea fowl	*Numida meleagris* spp.	Little egret	*Egretta garzetta*
Quail	*Coturnix coturnix japonica*	Cormorant	*Phalacrocorax carbo sinensis*
Pigeon	*Columba livia*	Bobwhite quail	*Colininus virginianus*
Pheasant	*Phasianus* spp. *Phasianus colchicus*	Peafowl	*Pavo cristatus*

Source: Landsborough Thomson (1964)

TABLE 2.11   Major agricultural fish

Herbivorous fish

Silver carp	*Hypophthalmichthys molitrix*
Sandkhol carp	*Thynnichthys sandkhol*
Grass carp	*Ctenopharyngodon idellus*
Striped mullet	*Mugil cephalus*
	Mugil capito
Golden gray mullet	*Mugil auratus*
Tilapia	*Tilapia* spp.
Milkfish	*Chanos chanos*
Ca ven	*Megalobrama bramula*
Tawes	*Barbus gonionotus*
Rohu	*Labeo rohita*
Mrigal	*Cirrhina mrigala*
Reba	*Cirrhina reba*
Gourami	*Osphronemus goramy*
Kissing gourami	*Helostoma temmincki*

Carnivorous fish

Catfish	*Pangasius* spp.
	Clarias spp.
Roach	*Rutilus rutilus*
Tench	*Tinca tinca*
Pike	*Esox lucius* — Northern Pike
Perch	*Anabas testudineus, Lateolabrax japonicus, Perca fluviatilis*
Bream	*Sparus auratus, Parabramis pekinensis*
Black or snail carp	*Mylopharyngodon* spp.
Ma lang yu	*Squaliobarbus curriculus*
Paddlefish	*Polyodon spathula*
Sheatfish	*Siluris glanis*
Bass	*Micropterus salmoides*
	Morone saxatilis
Trout	*Salmo* spp.
	Salvelinus spp.
Walleye	*Stizostedion vitreum*
Yellowtail	*Seriola quinqueradiata*
Bluegill	*Lepomis macrochirus*
Snakeheads	*Ophicephalus* spp.
Sleeper gobies	*Oxyeleotris* spp.
Loaches	*Cobitidae*
Eels	*Anguilla* spp.

TABLE 2.11 (*cont.*)

Sturgeon	*Acipenseridae*
Salmon	*Salmo* spp., *Thymallus* spp., *Salvelinus* spp., *Oncorhynchus* spp.
Grayling	*Thymallus thymallus*
Whitefish and Ciscoes	*Coregonidae*
Shad	*Alosa* spp., *Hilsa* spp.
Pompano	*Trachinotus carolinus*
Puffers	*Fugu* spp.
Porgys	*Chrysophrys major, Mylio macrocephalus*
Cod	*Gadus morhua*
Plaice	*Pleuronectes platessa*
Sole	*Solea solea*

Omnivorous fish

Carp	*Cirrhinus* spp., *Carassius* spp., *Cyprinus* spp., *Labeo* spp. *Thynnichthys* spp.
Big Head	*Aristichthys nobilis*
Belinka, Lampai	*Barbus* spp.
Tambra	*Labeobarbus tambroides*
Catla	*Catla catla*
Nagendram fish	*Osteochilus thomassi*
Buffalo fish	*Ictiobus* spp.
Catfish	*Ictalurus* spp., *Pylodictis* spp.
Clariids	*Clarias* spp.
Gouramis	*Trichogaster* spp.
Pejerrey	*Odonthestes basilichthys*
Smelts	*Osmeridae*
Ayu	*Plecoglossus altivelis*

Source: Bardach *et al.* (1972)

TABLE 2.12   Major agricultural invertebrates

Bees	*Apis mellifera*
	Apis indica
Clams	Nine species cultured in Japan
Hard clams or Quahogs	*Mercenaria mercenaria*
Soft clams	*Mya arenaria*
Cockles	*Anadara* spp.
Crabs	
Swimming crabs	*Portunidae*
Mud crabs	*Xanthidae*
Cancer crabs	*Cancridae*
Crayfishes	*Astacidae*
Cuttlefish	*Euprymna* spp.
	Sepia and *Sepiella* spp.
Lobsters	*Jasus* spp.
	Panulirus spp.
	Homarus spp.
Marine gastropods	Abalones, *Haliotis* spp.
	Murex trunculus
Mussels	*Mytilus* spp.
Oysters	*Crassostrea* spp.
	Pinctada spp.
	Ostrea spp.
Scallops	
Deep-sea scallop	*Patinopecten yessoensis*
Shrimps/Prawns	*Penaeus* spp.
	Macrobrachium spp.
Silkworms	*Bombyx mori*
Snails	
Land snails	*Helix* spp.
Sea snails	*Strombus* spp.
Sponges	
Squids	*Sepioteuthis lessoniana*

Source: Bardach *et al.* (1972)

TABLE 2.13 Major agricultural reptiles and amphibians

Frogs	
Bullfrog	*Rana catesbiana*
Greenfrog	*Rana clamitans*
Pickeral frog	*Rana palustris*
Leopard frog	*Rana pipiens*
	Rana hexadactyla
	Rana tigrina
Turtles	
Green turtle	*Chelonia mydas*
Hawkesbill turtle	*Evetmochelys imbricata*

Source: Bardach *et al.* (1972)

TABLE 2.14 Important agricultural plants

Family	Examples
Thallophyta	
algae	*Porphyra* spp. (Red algae)
	Undaria (Brown algae)
	Laminaria (kelp)
	Eucheuma spp.
fungi	cultivated mushroom (*Agaricus bisporus*)
	Shii-take (*Lentinus edodes*)
	Straw mushroom (*Volvariella volvacea*)
	Winter mushroom (*Flammulina velutipes*)
Pinaceae	Pines e.g. Scots pine, Cluster pine
	Spruces e.g. Sitka spruce
	Pseudotsuga spp. e.g. Douglas fir.
	Larches
Gramineae	Oat, millet, barley, rice, sugar cane, rye, sorghum, wheat, maize, etc.
Palmae	Coconut, oil palm, date
Araceae	Taro
Bromeliaceae	Pineapple

TABLE 2.14 (*cont.*)

Family	Examples
Liliaceae	Onion, shallot, garlic, asparagus
Amaryllidaceae	Sisal
Dioscoreaceae	Yam
Musaceae	Banana, hemp
Zingiberaceae	Turmeric, cardamom, ginger
Cannaceae	Queensland arrowroot
Marantaceae	West Indian arrowroot
Orchidaceae	Vanilla
Piperaceae	Pepper
Salicaceae	Poplars, e.g. Cottonwood
Juglandaceae	Pecan, walnut
Betulaceae	Hazelnut, filbert
Moraceae	Breadfruit, fig, hops, mulberry
Urticaceae	Ramie
Proteaceae	Macadamia nut
Polygonaceae	Buckwheat, rhubarb
Chenopodiaceae	Beet, spinach, chard
Basellaceae	Melloco
Caryophyllaceae	Carnation
Ranunculaceae	Aconite
Annonaceae	Soursop (papaw not true papaya)
Myristicaceae	Nutmeg
Lauraceae	Cinnamon, sweet bay, avocado

TABLE 2.14 (*cont.*)

Family	Examples
Papaveraceae	Opium poppy
Capparidaceae	Caper
Cruciferae	Mustard, cabbage, turnip, radish
Saxifragaceae	Gooseberry, black-currant, red-currant
Rosaceae	Quince, strawberry, apricot, peach, apple
Leguminosae	Groundnut, soybean, indigo, lentil, alfalfa, pea, beans, clover Acacias, e.g. wattle
Oxalidaceae	Oca
Linaceae	Flax
Rutaceae	Citrus fruit, kumquat
Burseraceae	Frankincense, pili nut, myrrh
Euphorbiaceae	Tung oil, rubber, cassava
Anacardiaceae	Cashew, mango
Aquifoliaceae	Maté
Aceraceae	Sugar maple
Vitaceae	Grapes
Tiliaceae	Jute
Malvaceae	Cotton, okra
Bombacaceae	Kapok
Sterculiaceae	Cola, cacao
Ternstroemiaceae	Tea
Passifloraceae	Granadilla
Caricaceae	Papaw or papaya

TABLE 2.14 (*cont.*)

Family	Examples
Punicaceae	Pomegranate
Lecythidaceae	Brazil nut
Myrtaceae	Allspice, guava, clove. eucalypts e.g. blue gum, red gum
Umbelliferae	Celery, carrot, parsnip, dill
Ericaceae	Cranberry, blueberry
Oleaceae	Olive
Loganiaceae	*Nux vomica*
Convolvulaceae	Sweet potato
Verbenaceae	Teak
Labiatae	Lavender, sweet marjoram, spearmint, thyme
Solanaceae	Red pepper, tomato, tobacco, aubergine, potato
Scrophulariaceae	Digitalis
Pedaliaceae	Sesame
Plantaginaceae	Psyllium
Rubiaceae	Ipecac, quinine, coffee
Valerianaceae	Valerian
Cucurbitaceae	Melon, cucumber, squash
Compositae	Endive, chicory, Jerusalem and globe artichoke, lettuce, salsify

Sources: Hill (1952); Logan (1967); Hayes (1977)

The atmospheric gases and some soil constituents would be better grouped together as environmental resources. Solar energy could be included here but it seems more useful to separate renewable and non-renewable resources (Table 2.16). Whether resources are regarded as renewable or not depends upon the time-scale: coal and natural gas are renewable over a very long period of time but energy in general is not. There is considerable confusion over whether energy can be used more than once and the literature abounds with statements that once it is used it is lost, dissipated as heat. But, of course, heat can be used and may result in other transfers of energy, so it is rarely quite so simple. The fact is that there are some unavoidable losses whenever energy is used and that eventually all of it is lost this way, but fossil fuels illustrate just how long "eventually" may be.

Similarly, it is often argued that energy cannot be recycled and that all

TABLE 2.15 Examples of other organisms essential to many agricultural operations

Agricultural operation	Organism	Examples
Maintenance of a satisfactory soil structure	Earthworms	*Lumbricus* spp. *Allolobophora* spp.
Dispersal and decomposition of dead plant and animal tissue and faeces	Small mammals, birds, insects, micro-organisms	Voles, crows, flies, dung beetles, bacteria
Nitrogen fixation	Free-living and symbiotic micro-organisms	Blue–green algae *Spirithum lipoferum* *Rhizobium* spp.
Pollination	Insects, birds and bats	Honey bee — *Apis mellifera* Bumblebees — *Bombis* spp. Many other bees, wasps and other insects
Functioning of the digestive tract of agricultural animals	Gut flora and fauna	*Ruminococcus albus* *Streptococcus bovis* Coliforms, Protozoa
Biological control of weeds and pests	Parasite or predator of pest	*Encarsia* (on whitefly); cochineal scale and coreid bugs (on prickly pear)
Food processing	Lactic acid-producing organisms and yeasts	*Streptococcus lactis* *Streptococcus cremaris* *Lactobacillus bulgaricus* *Saccharomyces* spp.

other resources can. This, too, is an oversimplification. Solar energy fixed by plants can be used by animals, which may be consumed by other animals, or which may die and provide an energy substrate for all kinds of organisms, including plants (as bacteria and fungi). The energy is certainly cycling in some senses but is not used again for photosynthesis, although it is conceivable that it would contribute heat to plant growth, which is one of the functions of solar energy.

Complications of this kind make it extremely difficult to determine the cost of some important resources.

TABLE 2.16 The main environmental resources for agriculture

Renewable	Non-renewable (in the short term)
Gases	
O_2	Coal
CO_2	Natural gas
N	Oil and its products
Water	Peat
Phosphates	
Potash	
Solar energy	
Wind	
Water power	
Soil	
Lime	

The Cost of Resources

Cost may reflect scarcity or demand as well as the cost of production or extraction, and the latter often have to bear the costs of finding the resources in the first place.

Some resources are limited in quantity. Land is an obvious example, although land is sometimes created, but the total surface of the earth is of fixed area. Solar radiation is also a fixed quantity per day but that quantity is enormous and may be expected to continue daily for the forseeable future.

The increasing dependence of agriculture on support energy is one reason for examining the extent of these resources but more important is the rate at which they are being consumed by industry and domestically. Fossil fuels are present in enormous quantities but (a) the current rate of use is high and still increasing and (b) much of the stored energy is difficult, dangerous and costly to obtain. Indeed, the *energy* costs are becoming very high and this indicates an upper limit: when the energy costs of exploration, extraction, processing, storage and delivery exceed the energy supplied, the activity is both profitless and pointless. Furthermore, these energy costs have to include *all* the inputs required to manufacture the drilling platforms and the ships that service them and all the other equipment used in all stages of the process, up to the point of delivery to the customer.

The same argument applies to renewable resources. The energy costs of recovering and recycling minerals, for example, could become very high and effectively render some of them non-renewable. This applies to the recycling of water, too, except that solar energy already does this on an enormous scale and could be harnessed to do more.

For many reasons, therefore, although particular resources (e.g. phosphate) may represent severe limiting factors at certain times and in particular places, the major limitation imposed on all human activity appears to be the energy supply. If better use could be made of solar radiation or new sources of energy discovered or known sources (e.g. nuclear fission or fusion) developed safely to a point where we could manage without fossil fuels, there would be no great problem. It is difficult, at the moment, however, to see how fossil fuels can be replaced by substitutes in the time left before they are exhausted. Figure 2.2 shows the rate at which they are currently being used and Table 2.17 summarizes the probable reserves.

There are strong indications, therefore, that the use of support energy may have to be restricted, or will be restricted by increasing prices, and that the efficiency with which it is used will be of paramount importance. This will naturally figure in many of the efficiency calculations, given in subsequent chapters, but it is a good example of the difficulties and dangers of looking at one efficiency ratio by itself. Clearly, it is also desirable to increase the efficiency with which solar radiation is used and it is unlikely that one would wish to examine the use of support energy without considering its interactions with the use of solar energy. Just which resources are the most relevant for the calculation of efficiency varies with the environment and the extent to which plants and animals are adapted to it. For this reason, this topic is dealt with first in each of the next two parts of the book.

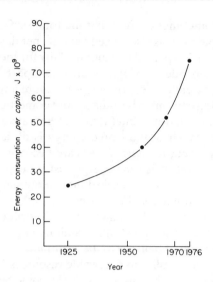

FIG. 2.2   World consumption of fossil fuels on a *per capita* basis

TABLE 2.17   Probable world reserves of recoverable energy (after King Hubbert, 1971)

	Quantity of reserve	Energy content of reserve MJ × 10¹⁵
Coal and lignite	7.6×10^{12}t	201.2
Natural gas	1000×10^{12}ft^3	10.6
Petroleum liquids	2000×10^9 barrels	11.7
Tar-sand oil	300×10^9 barrels	1.8
Shale oil	190×10^9 barrels	1.2

References

Bardach, J. E., Ryther, J. H. and McLarney, W. O. (1972). "Aquaculture. The Farming and Husbandry of Freshwater and Marine Organisms", Wiley, New York.

Department of Employment (1973). Department of Employment Gazette. Her Majesty's Stationery Office.

F.A.O. (1973). State of Food and Agriculture. F.A.O., Rome.

Hayes, W. A. (1977). Mushrooms for every home. Paper presented to the B.A. Section M — Agriculture, Sept. 1977.

Hill, A. F. (1952). "Economic Botany", 2nd edn, McGraw-Hill, New York.

Kilkenny, J. B. (1976). Economic significance of wool in meat production systems. *Agric. Syst.* **1**, 313–320.

King Hubbert, M. (1971). The energy resources of the earth. *Scientific American* **225(3)**, 60–70.

Landsborough Thomson, A. (1964). "A New Dictionary of Birds", Nelson, London.

Logan, W. E. M. (1967). FAO world symposium on man-made forests and their industrial importance. I. Policy. *Unasylva* **21(3–4)**, Nos. 86–87, 8–22.

Masefield, G. B., Wallis, M., Harrison, S. G. and Nicholson, B. E. (1969). "The Oxford Book of Food Plants", Oxford University Press, London.

McKay, D. H. and Ward, A. (1962). The structure of the Australian sheep industry. Chap. 22 *in* "The Simple Fleece" (Ed. A. Barnard), Melbourne University Press, Australia.

The Public Ledger (1973). U.K. Publications Ltd., London.

Rao, K. K. P. N. (1976). Food Consumption and Planning. Vol. 5 International Encyclopaedia of Food and Nutrition. Pergamon Press, Oxford.

U.N. Economic Commission for Europe, F.A.O. (1974). Prices of Agricultural Products and Selected Inputs in Europe 1972/73. U.N., New York.

Part II
Efficiency in Crop Production

3

ADAPTATION TO THE ENVIRONMENT

Since survival of an organism depends upon some degree of adaptation to the environment, it seems obvious that those species that are well adapted will be the most efficient. In some senses, especially in terms of the survival of the species in the natural environment, this is true, but in an agricultural context the environment is never wholly natural. So it is that some plant species that are well equipped to survive are not very productive: their low productivity may be closely related to their ability to survive and may even be part of their adaptation.

What is really required agriculturally, therefore, is a plant species that is well adapted to the (mainly) unalterable features of the environment (such as temperature, day-length and light intensity) but responsive to improvements in the controllable features (such as water and nutrient supply and soil pH status). Even where some of these features *could* be controlled, of course, it does not follow that this would be economically feasible, so there is great variation in the degree of adaptation required. Furthermore, agricultural environments include a wide range of protected situations, from greenhouses to shelterbelts (Stott and Belcher, 1978). The latter may be trees but hedges and walls are also used as barriers to wind, generally influencing crop growth for a distance up to 30 times the height of the barrier (Gloyne, 1976).

The limitations imposed on efficiency assessments are fairly obvious and no-one would probably wish to compare turnip production with that of lichens in the Arctic, or to compare the relative efficiency with which rice and wheat use boggy land.

This mainly illustrates again the fact that one species is not likely to be more efficient than another at doing all things in all circumstances, and that the degree of adaptation to the environment may greatly affect the answers. Equally, however, it is important to recognize that it may be desirable to compare the productivity of wheat and rice, where, for example, there is a

41

choice as to whether land should be flooded or not. Indeed, it may then be sensible to compare wheat and fish production.

Adaptation to the environment is not then something that can be dealt with quite separately but needs to be borne in mind in considering different sorts of plants and animals. This is important to recognize because a number of interesting questions are influenced by it. For instance, whether trees are more efficient than grass in the use of water, or whether perennials are more efficient than annuals in the use of sunlight: they are still interesting questions but the answers will vary with the environment considered.

The aim of this chapter is simply to give a picture of the main environmental features and the most important ways in which crop plants are adapted to them. One of the biggest differences between environments is between terrestrial and aquatic plants.

Aquatic Plants

These may be either freshwater or marine, or adapted to brackish water. Very few have been used agriculturally, which seems a little surprising when plants such as water hyacinth (*Eichhornia crassipes*) (Little, 1968) grow so vigorously that they create a problem by physically blocking tropical waterways (Wolverton and McDonald, 1976). Many aquatic plants are not attached to the bottom and thus present few harvesting problems and they do not require strong, fibrous supporting structures of the kind that render many terrestrial plants unsuitable for human food. There are difficulties in ever achieving a high concentration of nutrients in large bodies of water and fertilizer application is virtually restricted to shallows and enclosed lakes and ponds (de Wit, 1967).

Seaweeds have been used but rarely cultivated on any scale and would not normally be classed as agricultural plants yet, although an experimental farm unit was established off the tip of San Clemente Island, California (Wilcox, 1975), using *Macrocystis pyrifera* and feeding the product to sheep, snails, abalone and fish. An exception to the general rule is the farming of the red-weed *Porphyra* spp., which have been cultivated along the coastline of Japan, especially in Tokyo Bay, for some 300 years (Korringa, 1976). Porphyra (Japanese "nori") may yield up to 750 kg ha^{-1} during the six to eight months growing season (Bardach *et al.*, 1972). The brown alga *Undaria* is also cultivated in Japan, yielding up to 10 kg of wet weed per 1 m of cultivating rope, and *Laminaria* is grown in both Japan and China, sometimes involving dynamite to improve the substrata and control harmful weeds. The energetic efficiency of this process has not been calculated. In the Phillipines, the most successful seaweed culture is of *Eucheuma* sp., with yields of *c.* 13 tonnes ha^{-1} (on a DM basis).

However, some 2.4 million tonnes (wet weight) of seaweed were harvested in 1973, with an annual value currently (1976) estimated at about U.S. $1000 million, using a wide variety of species and producing a wide range of products (Naylor, 1976). In principle, this may not be so different from the utilisation of permanent pasture in agriculture.

Rice and other plants that tolerate or prefer very wet and often flooded land are not really aquatic, although, as in the case of rice, they may benefit from being associated with aquatic blue–green algae (*Cyanophyceae*) (Grist, 1975; Nutman, 1971). Watercress (*Nasturtium officinale*) is grown regularly in flowing water (Bardach *et al.*, 1972).

In general, there is a tendency for aquatic plants to be microscopic and to be consumed by microscopic animals (consumed in turn by, largely carnivorous, fish).

Terrestrial Plants

Land plants, by contrast, often have elaborate root systems, for anchorage and to obtain both water and nutrients, and highly structured above-ground parts, in order to stand erect and compete for light. They therefore often grow quite large and may contain some very strong and hard elements. There is no aquatic equivalent of the tree, although there is no shortage of microscopic terrestrial plants. In general, the problems of land plants (including dispersal, for example) have resulted in a great variety of species and it is sometimes argued (see Janick *et al.*, 1969) that we have made agricultural use of very few (less than 0.4% of the available higher plants). It may be that other species could be used agriculturally but they tend to appear less useful in comparison to the highly selected varieties of the species now in use.

Light

Several aspects of incident light are important for plants. Most require direct solar radiation (in the 0.4–0.7 μm wave region of the spectrum) to provide the energy for photosynthesis. Exceptions include many bacteria and fungi, deriving their energy from organic substrates.

Plant species are adapted to both light intensity and day-length and both of these features may greatly influence aspects of efficiency. A good illustration of this is provided by the differences in respiration pathways characteristic of the C_3 and C_4 plants (see Table 3.1), enabling the latter to make more efficient use of very high light intensities.

Plants vary greatly in their height and above-ground structure and thus in

TABLE 3.1   Differences in the respiration pathways characteristic of the C_3 and C_4 plants

---

C_3 Leaves

   1. The first product of photosynthetic carbon fixation is a three-carbon carboxylic acid, phosphoglyceric acid.

   2. The CO_2 compensation point is high (around 50 ppm).

   3. Net photosynthesis is inhibited by atmospheres containing more than about 1% oxygen.

   4. There is no "Kranz-type" leaf anatomy.

C_4 Leaves

   1. The first product of photosynthetic carbon fixation is a four-carbon dicarboxylic acid, oxaloacetic acid.

   2. The CO_2 compensation point is close to zero.

   3. Oxygen has no major effect on the rate of net photosynthesis.

   4. Leaves do possess the "Kranz-type" anatomy in which the well-developed bundle-sheath cells are rich with organelles, including large chloroplasts, mitochondria and peroxisomes.

Some C_3 Plants	Some C_4 Plants
Sugar beet, *Beta vulgaris*	Sugar cane, *Saccharinum* spp.
Alfalfa, *Medicago* spp.	Maize, *Zea mays*
Soyabean, *Glycine max.*	Sorghum, Sudangrass, *Sorghum* spp.
Wheat, *Triticum vulgare*	Bermuda and Stargrass, *Cynodon* spp.
Oil palm, *Elaeis quincensis*	Napiergrass, *Pennisetum purpureum*
Cassava, *Manihot esculenta*	
Potato, *Solanum tuberosum*	
Rice, *Oryza sativa*	
Ryegrass, *Lolium* spp.	

Sources: Gifford (1974); Loomis and Gerakis (1975)

their efficiency as light-receiving surfaces. In addition, they vary in plant density and it is very often the plant population that has to be dealt with when considering areas of land or water. Furthermore, the relevant situation may involve populations of mixed species and mixed ages of individuals.

This adds to the complexity but is a major characteristic of natural eco-systems. Agricultural systems, by contrast, often aim at much simpler mixtures or even single-species populations, in order to optimize manage-ment more easily.

A very important question remains, in many agricultural situations, as to whether more mixed cropping would make better use of resources, or in what circumstances this would be so. Table 3.2 illustrates the effects that have been found to occur on occasion.

TABLE 3.2   Examples of mixtures giving substantially higher yields, or greater financial returns than pure stands of the component species (from Willey and Roberts, 1976)

Reference[b]	Species	Yield increase, or greater financial return[a]
(1)	Long duration sorghum, intercropped with:	
	(i) Maize followed by cowpea (1969)	70%[a]
	(ii) Sorghum followed by cowpea (1970)	80%[a]
(2)	Maize/Rice	33%
	Maize/Cassava	15%
	Rice/Cassava	35%
	Maize/Rice/Cassava	62%
(3)	Setaria millet/Pigeon pea	64%[a]
	Pearl millet/Pigeon pea	40%[a]
	Sorghum/Pigeon pea	23%[a]
	Green maize/Sorghum/Pigeon pea	90%[a]
(4)	Maize/Phaseolus beans	38%
(5)	Sorghum/Phaseolus beans	55%
(6)	Maize/Soya bean	22%
(7)	Fodder radish/Sunflower	26%

[b](1) Andrews (1972)
(2) Int. Rice Res. Inst. (1975)
(3) Krantz et al. (1976)
(4) Willey and Osiru (1972)
(5) Osiru and Willey (1972)
(6) Osiru (1974)
(7) Lakhani (1976)

Temperature

Plant growth is generally very slow below 7°C and very few plant species are adapted to temperatures above 38°C. However, the actual temperature to which plants are exposed varies with altitude, latitude, season, cloud cover, shade from other plants, topography and many other features.

Quite apart from relationships between growth rate and ambient temperature, there are also aspects such as resistance to frost, response to variations in day and night temperature and effects of temperature on opening of flowers (and activity of pollinating insects) and of seed and fruit ripening. Seasonal patterns of temperature may therefore be of consequence; but these may interact with related environmental features, especially water supply.

Water Supply

A plentiful supply of water is essential for high production by plants (Begg and Turner, 1976), so many adaptations to water shortage may be associated with low growth rates. However, water shortage may take many forms and adaptation may allow plants to survive periodic shortages, by reduction in water loss or by some form of storage, or may be associated with additional methods of obtaining water. Deep or extensive rooting may enable plants to obtain enough water, even during a general shortage. This will only result in growth and production if the supply of nutrients is also adequate and there are often interactions between water and mineral supplies. This is most noticeable in relation to soil nitrogen, since this may be present in greatest concentrations in those parts of the soil that dry out first in a drought. Plants that tap deep reserves of water may therefore lack nitrogen, and legumes may have great advantages in such situations, provided that their nitrogen-fixing nodules remain functional.

Nutrient Supply

Nitrogen is a good example of a nutrient which may be present in very variable amounts and concentrations, to which plants may be well adapted. This is very noticeable with legumes but the list of plant species that benefit from some kind of association with nitrogen-fixing microorganisms is considerable (Table 3.3) and the contribution of free-living organisms has yet to be fully determined (Mulder, 1975; Dobereiner and Day, 1975; Day et al., 1975; Ruinen, 1975; Henriksson et al., 1975; Venkataraman, 1975).

TABLE 3.3 Plant species that benefit from some kind of association with nitrogen-fixing micro-organisms (after Nutman, 1971)

Angiospermae	
Dicotyledons	
Leguminosae	*Rhizobium* spp.
Rosaceae	
Eleagnaceae	
Rhamnaceae	*Actinomyces*
Ericaceae	*Streptomyces*
	or
Coriariceae	*Nocardia?*
Myricaceae	
Betulaceae	
Casuarinaceae	
Rubiaceae	*Klebsiella*
Myrsinaceae	*Xanthomonas*
Halorrhagaceae	*Nostoc punctiforme*
Monocotyledons	
Dioscoreaceae	bacterial
Graminae	bacterial e.g. *Azotobacter paspali*
Gymnospermae	
Cycadaceae	*Anabaena, Nostoc*
Pteridophyta	
Filicinae	*Anabaena azollae*
Bryophyta	
Hepaticae	*Nostoc*
Lichens	various blue-green algae

Other nutrients, especially potash and phosphate, also vary in concentration in the soil and plants vary greatly in how extensive their root systems are and thus in the volume and depth of soil they explore.

Plant species also differ in the magnitude of their response to additional nutrients: this is not always a simple matter of growth in response to fertilizers, of course, since by no means all plant products are the simple result of growth and product quality may be adversely affected by, for example, too much nitrogenous fertilizer.

Trace elements are often important in determining the occurrence of plant species but, agriculturally, are relatively easy to supply, because the quantities involved are small.

Pests and Disease

All plants are open to attack by a range of pests and diseases, but some are much more vulnerable than others. Since the incidence of both pests and disease may be influenced by features such as plant density, and the susceptibility of the plant may be affected by growth rate and nutrient supply, the effect on efficiency of land use may be very complicated. Clearly, measurement of production needs to be accompanied by measurement of pest and disease status in order to state what an efficiency assessment represents. Damage may occur internally or externally and to any part of the plant, above or below ground, and it is not always very obvious. An efficient plant may be one that is vulnerable to few pests and diseases (because it is tough, hairy or poisonous, for example), one that can grow so vigorously that it replaces lost or damaged parts, or one that can continue to function although much damaged. Agriculturally valuable plants are rather more likely to come from the middle group, especially if they are for direct food consumption, but it is difficult to generalize usefully across such a wide range.

It is an interesting proposition that more efficient agricultural plants might be produced by breeding for resistance to pests, without reduction in, for example, food quality: this is done on a considerable scale already for disease resistance.

Plants are also vulnerable to damage from larger animals, including mice, voles, rats, squirrels and all the major herbivores, though these are rarely called pests.

Plant Form

Small pests, such as insects, are not unaffected by plant form and structure but they can occur on any kind of plant. This is less true of the larger animals, and trees, for example, though browsed by some tall animals, are grazed by relatively few. This argument obviously depends upon the precise definition of terms like grazing and browsing but the point here is a very simple one, that plant form is one kind of adaptation to attack and damage by larger herbivores (although there are, in fact, quite a number of climbing herbivores, e.g. koala bear, panda, gorilla) and omnivores. The agricultural significance of adaptation by way of form and structure is limited by the relative ease with which large animals can be eliminated from the agricultural scene.

However, form and structure also play a part in adaptation to other environmental features. Light has already been referred to in this context and it is rather obvious in this case how the importance of structures varies with the particular mixture of species: similar competitive situations arise in relation to root structures and the supply of nutrients. Above-ground structure may also affect gaseous exchanges and evapo–transpiration by placing the foliage in a different, and sometimes more turbulent, microclimate.

Agricultural Significance of Adaptation

There are clear senses in which (a) only plants that are adapted to an environment will grow in it and (b) the better adapted the plant the more agriculturally useful and productive it will be. However, agricultural practices include the use of a whole range of cultural, chemical and even biological agents that can render some adaptations unnecessary and agriculture may radically change much of the environment in which plants are grown. Horticulture, for example, commonly involves the almost total control of light, temperature, rooting medium, nutrient and water supply and even the concentration of CO_2.

The calculation of an efficiency ratio nevertheless must relate to a specified environment and it must be expected that adaptations will influence efficiency. Furthermore, specific adaptations will clearly affect particular efficiencies and the many possible combinations of different adaptations may result in different agricultural efficiencies in successive periods (of weeks, seasons or years). This raises another aspect of agriculture in variable weather conditions: the objective may have to do with consistency of result, rather than large variations including both very good and very poor results. So it is possible to have adapted mixtures of species, in the sense that if the weather is adverse for one, it may favour another and no year is ever a total disaster.

Thus, although the advantage of adaptation to environment seems obvious, it is less clear how this can be achieved for very variable environments. One advantage of the agricultural activity is that considerable control can often be exercised in the direction of reduced variation, especially in such things as water and nutrient supply.

References

Andrews, D. J. (1972). Intercropping with sorghum in Nigeria. *Expl. Agric.* **8**, 139.

Bardach, J. E., Ryther, J. H. and McLarney, W. O. (1972). "Aquaculture", Wiley-Interscience, New York.

Begg, J. E. and Turner, N. C. (1976). Crop water deficits. *Adv. Agronomy* **28**, 161–217.

Day, J. M., Harris, D., Dart, P. J. and Van Berkum, P. (1975). The Broadbalk experiment. An investigation of nitrogen gains from non-symbiotic nitrogen fixation. *In* "Nitrogen Fixation by Free-living Micro-organisms", (Ed. W. D. P. Stewart), Cambridge University Press, Cambridge.

Dobereiner, J. and Day, J. M. (1975). Nitrogen fixation in the rhizosphere of tropical grasses. *In* "Nitrogen Fixation by Free-living Micro-organisms", (Ed. W. D. P. Stewart), Cambridge University Press, Cambridge.

Gifford, R. M. (1974). A comparison of potential photosynthesis, productivity and yield of plant species with differing photosynthetic metabolism. *Aust. J. Plant Physiol.* **1**, 107–117.

Gloyne, R. W. (1976). Shelter in agriculture, forestry and horticulture — a review of some recent work and trends. *ADAS Q. Rev.* **21**, 197–207.

Grist, D. H. (1975). "Rice", 5th edn, Longmans, London.

Henriksson, E., Henriksson, L. E. and Dasilva, E. J. (1975). A comparison of nitrogen fixation by algae of temperate and tropical soils. *In* "Nitrogen Fixation by Free-living Micro-organisms", (Ed. W. D. P. Stewart), Cambridge University Press, Cambridge.

International Rice Research Institute (1975). *Ann. Rep.* 1974.

Janick, J., Schery, R. W., Woods, F. W. and Ruttan, V. W. (1969). "Plant Science", W. H. Freeman, San Francisco.

Korringa, P. (1976). "Farming Marine Organisms Low in the Food Chain", Elsevier, Amsterdam and Oxford.

Krantz, B. A., Virmani, S. M., Sardar Singh and Rao, M. R. (1976). Intercropping for increased and more stable agricultural production in the semi-arid tropics. Symposium, on Intercropping, Morogoro, Tanzania, May 1976.

Lakhani, D. A. (1976). A crop physiological study of mixtures of sunflower and fodder radish. Ph.D. Thesis, University of Reading, U.K.

Little, E. C. S. (1968). "Handbook of Utilisation of Aquatic Plants", (Ed. E. C. S. Little), F.A.O. Rome.

Loomis, R. S. and Gerakis, P. A. (1975). Productivity of agricultural ecosystems. *In* "Photosynthesis and Productivity in Different Environments". (Ed. J. P. Cooper), I.B.P. 3. Cambridge University Press, Cambridge.

Mulder, E. G. (1975). Physiology and ecology of free-living nitrogen-fixing bacteria. *In* "Nitrogen Fixation by Free-living Micro-organisms", (Ed. W. D. P. Stewart), Cambridge University Press, Cambridge.

Naylor, J. (1976). Production, trade and utilization of seaweeds and seaweed products. F.A.O. Fisheries Tech. Paper No. 159, F.A.O., Rome.

Nutman, P. S. (1971). Perspectives in biological nitrogen fixation. *Sci. Prog. Oxf.* **59**, 55–74.

Osiru, D. S. O. (1974). Physiological studies of some annual crop mixtures. Ph.D. Thesis, University of Makerere, Uganda.

Osiru, D. S. O. and Willey, R. W. (1972). Studies on mixtures of dwarf sorghum and beans (*Phaseolus vulgaris*) with particular reference to plant population. *J. Agric. Sci., Camb.* **79**, 531.

Ruinen, J. (1975). Nitrogen fixation in the phyllosphere. *In* "Nitrogen Fixation by Free-living Micro-organisms", (Ed. W. D. P. Stewart), Cambridge University Press, Cambridge.

Stott, K. G. and Belcher, A. R. (1978). Living windbreaks: a review of work at Long Ashton. Long Ashton Research Station, Report 1977, 204–218.

Venkataraman, G. S. (1975). The role of blue-green algae in tropical rice cultivation. *In* "Nitrogen Fixation by Free-living Micro-organisms", (Ed. W. D. P. Stewart), Cambridge University Press, Cambridge.

Wilcox, H. A. (1975). The ocean food and energy farm project. Paper presented to American Association for the Advancement of Science Jan 1975 and reported in *Simulation* (1975) **25**(6), and **26**(1).

Willey, R. W. and Osiru, D. S. O. (1972). Studies on mixtures of maize and beans (*Phaseolus vulgaris*) with particular reference to plant population. *J. Agric. Sci., Camb.,* **79**, 517.

Willey, R. W. and Roberts, E. H. (1976). Mixed cropping. *In* "Solar Energy in Agriculture", *Proc. Joint Conf.* University of Reading, and UK–ISES Sept. 1976.

Wit, C. T. de (1967). Photosynthesis: its relationship to over-population. *In* Harvesting the sun (Eds. A. San Pietro, F. Greer and T. J. Army). Academic Press, New York, London.

Wolverton, W. and McDonald, R. C. (1976). Don't waste waterweeds. *New Scientist* **72** (1013), 318–320.

4

THE SUPPLY OF NUTRIENTS

The main difference between the nutrient supply of plants in a natural ecosystem and that of crop plants in agricultural systems is that the latter involve removal of nutrients in harvested material. The quantities may be substantial where foods are produced for direct human consumption or where industrial raw materials are harvested: the amounts actually removed in animal production systems are usually much less, depending on their intensity and whether excreta are recycled or not. The quantities involved are illustrated in Table 4.1.

These quantities represent the minimum that have to be replaced as fertilizer, but there are often other sources of loss, as in drainage water and gaseous losses to the atmosphere, and some minerals (e.g. phosphate) may have to be added in greater quantities than those removed, because a proportion becomes "locked up" and unavailable to plants (Larsen, 1973).

In some soils, quantities added and removed are very large in relation to those naturally present in the soil. In very fertile soils, however, this may not be so: indeed soil may contain vast amounts of nitrogen; for example, up to 5000 kg ha^{-1} in the top six inches of English soils (Cooke, 1967). In some situations, there may be considerable advantages in such reserves: certainly, dependence on current fertilizer applications is risky when rainfall is inadequate, unless irrigation is available. The supply of nutrients is not only dependent on water supply but also on soil structure, the activity of soil flora and fauna and rates of fixation and release of elements. Considering the importance of the soil, there is often a surprising ignorance of the vital processes going on within it, in agricultural systems. It is, of course, difficult to see these processes and, in many cases, anything at all below the surface. The structure of the visible soil, especially in seed-beds, has received attention, but much less is known of the living organisms and their roles: similarly, much more is known about the above-ground parts of plants than about their roots.

52

TABLE 4.1 Quantities of nutrients removed in harvested crops

	Yield per ha per yr		Nutrients in yield kg ha^{-1}					
			N	P	K	Ca	Mg	S
Crop products								
Wheat grain and straw	4	tonnes grain	80	12	40	10	5	20
Beans (grain)	2.5	tonnes	110	15	50	20	5	25
Potatoes	50	tonnes tubers	180	25	200	10	15	20
Maize	13	tonnes	360	52	230	—	75	50
Rice	7	tonnes	150	25	17	—	17	20
Soybeans	3	tonnes	210	25	110	—	14	11
Animal products								
Cows' milk[a] (from grass alone)	6.7	tonnes	35	6	11	8	0.9	2
Grass to produce[b] that milk	9	tonnes DM	225	27	225	65	18	13.5
Hens' eggs[c]	12 000 eggs 696 kg		13	1.5	0.1	0.4	0.08	1.2
Feed for hens to produce these eggs[d]								
Barley	1 943 kg DM		33	5.8	14.5	4.9	1.9	7.8
Beans	894 kg DM		39	5.4	18	7	1.8	9
Total			72	11	33	12	4	17
Beef (single suckled) mainly grass fed	360 kg carcase		12	0.7	1.3	0.03	0.09	0.8

Sources include: Cooke (1972); McCance and Widdowson (1960)

[a]Nutrients removed in the milk alone: no allowance has been made for nutrients retained by the cow for growth and reproduction or lost to atmosphere as methane or by leaching from excreta. All faeces and urine returned to the pasture.

[b]If cows were zero-grazed and faeces and urine not returned to the pasture these quantities of nutrients would be removed.

[c]Fifty hens laying 240 eggs per year.

[d]Supplied diet to produce feed for 50 hens for a year.

The difficulty is that an efficiency assessment, like any other assessment, has only limited value if the system examined cannot be described and characterized. Without such knowledge, there is no rational basis for change and improvement and no basis for generalisation to other, sufficiently similar systems, since one cannot say what a similar system is.

This is well illustrated by recent findings relating to the response of grassland to applications of nitrogenous fertilizer (see Table 4.2 and Fig. 4.1). Responses varied from one site to another for no clear reason and the efficiency of nitrogen utilisation for herbage growth also varied considerably. Without further detailed descriptions of the systems studied, and their environments, however, it is not possible to make practical recommendations. This is a serious matter when it concerns a crop that is amongst the most demanding of nitrogen, itself one of the most important agricultural plant nutrients.

TABLE 4.2 Response of grassland to nitrogenous fertilizer

| Geographical region in England and Wales | N applied (kg per ha per year) | | | | |
| | 0 | 150 | 300 | 450 | 600 |
	Annual DM yield (t ha^{-1})				
North-east	3.19	4.87	10.17	11.33	11.49
South-east	1.74	4.98	7.60	9.27	9.73
Midlands	3.02	7.21	10.20	11.56	12.15
South-west (incl. S. Wales)	1.89	5.29	9.78	11.93	13.29
Wales	4.30	8.25	11.67	13.35	13.06

Data from: National Grassland Manuring Experiments. Regional means for four harvest years (after Morrison, 1975).

Nitrogen

This element has already been referred to (see also Chapter 3) but is worth further consideration in relation to efficiency.

Applications of nitrogenous fertilizer are required in very different quantities for different crops and conditions (Table 4.3) and will generally result in a curvilinear response in terms of growth (illustrated in Fig. 4.2). Efficiency of use of land (and solar radiation) will therefore increase with rate of nitrogen application (Tatchell, 1976) in a rather similar fashion and efficiency of protein production per unit of nitrogen applied will tend to decline (Fig. 4.3).

All such responses must depend upon other nutrients (and water) being present in adequate amounts.

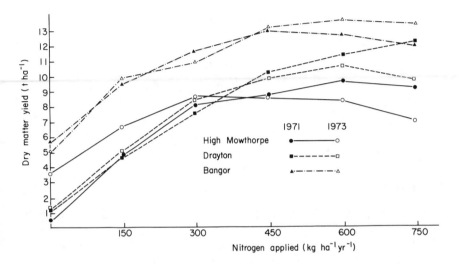

FIG. 4.1 Response of grassland at different sites to nitrogen applications (after Morrison and Jackson, 1976)

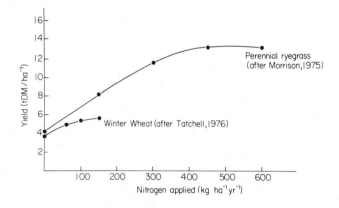

FIG. 4.2 Response of different crops to applications of nitrogen (after Morrison, 1975 and Tatchell, 1976)

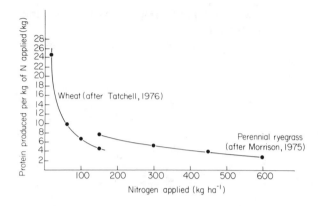

FIG. 4.3   Efficiency of protein production per unit of nitrogen applied (after Morrison, 1975 and Tatchell, 1976)

Potash

This is one of the major nutrients and is required in substantial quantities by many crops (Table 4.3).

Whereas nitrogen is freely available in the atmosphere, although there may be a high cost in fixing it non-biologically, potash has to be mined. Fortunately, the world contains large, fairly readily available and well distributed reserves. This is not the case with phosphates.

TABLE 4.3   Fertilizer requirements for different crops (kg ha^{-1})

Crop	Country	N	P_2N_5	K_2O
Wheat	UK	0–75	38	38
	Europe	50–150	40–80	40–80
	Australia	40–80	—	—
Maize	UK	100	75	75
	Europe	100–150	60–90	60–90
	USA	30–300	—	—
Field beans	UK	0–25	0–38	0–50
	Europe	0–20	50–75	50–75
Soybeans	USA	0	34–67	56–84
Potatoes (main crop)	UK	75–150	163–250	125–250
	Europe	120–200	120–220	150–250
Rice	Various countries	30–160	20–90	25–80
Sugar cane	Hawaii	up to 400	280	400–450
	Various countries	56–220	60–150	60–210

Sources: Eddowes (1976); Geus (1973); M.A.F.F. (1973); Martin et al. (1976); Purse-glove (1972)

Phosphate

Phosphates also have to be mined, unless they come from guano (massive deposits of sea-bird droppings), but world reserves tend to be concentrated in a limited number of enormous deposits, in U.S.A., Morocco and Russia (F.A.O., 1975).

Plants need substantial quantities (see Table 4.3) but availability is greatly influenced by soil pH.

Soil pH

The acidity of the soil influences what crops can be grown (see Table 4.4) and affects soil flora and fauna, so much so that virtually no decomposition of plant remains occurs below a pH of about 4.0, resulting in the formation, for example, of acid peat.

Massive applications of lime are used to alter soil pH: it can be applied in a variety of ways, including by air to otherwise inaccessible areas.

TABLE 4.4   Soil pH requirement for different crops

Crop	Optimum	Tolerable range	pH below which crop growth is noticeably restricted on mineral soils
Wheat	6.0–7.5	5.5–8.0	5.4
Maize	6.0–7.5	5.5–8.0	5.5
Field beans	—	—	6.0
Soyabeans	5.5–7.0	—	—
Potatoes	5.0–5.5	4.5–7.0	4.9
Rice	5.5–6.5	4.5–9.0	—
Sugar cane	—	5–8	4.5

Sources: Geus (1973); M.A.F.F. (1973); Martin *et al.* (1976); Purseglove (1972)

Animal Manures

Where animals graze, their excreta is usually returned to the pasture, but this can be very patchy, partly because this is the way it is deposited and partly because of the formation of "night camps" and other results of animal behaviour patterns.

The manure of housed or penned animals, often mixed with their bedding, has traditionally been used as a fertilizer in many countries and as a fuel or as building material in others.

The nutrients contained in animal excreta are indicated in Table 4.5 and the total quantities produced are given in Table 4.6.

When considering recycling of this kind, however, it has to be remembered that it does involve costs, both monetary and of energy, and is not without losses.

Recycling of Nutrients

Animal manure is an obvious source of nutrients for recycling but it should be recognized that plant material senesces and dies continuously throughout the life of the plant and is generally recycled by decomposition. In one sense, the main difference between this and grazing by large herbivores is simply the greater redistribution of nutrients due to the patterns of animal excretion.

The proportion of the whole crop that is harvested varies enormously with the species, the product and the environment; but the total considered at any one harvest time is, of course, only the *net* production. There are always growth and development processes that are essential, and require nutrients, that do not form part of the product and may have been lost long before harvest. Efficiency calculations of nutrient use therefore have to distinguish between the nutrients supplied that finish up in the product, those that are used in its production and those that fail to reach the plant at all or are taken up unnecessarily. The last, often referred to as "luxury uptake", occurs with nitrogen and potash, for example, and the content of these elements in the plant may continue to rise long after there is any growth response to their presence in the soil.

Recycling occurs automatically in relatively closed systems and therefore may be regarded as occurring in the world as a whole (Söderlund and Svensson, 1976). Agricultural activities are often seen as extractive but they cannot continue for long if essential minerals are not replaced. In intensive agricultural systems, however, this replacement generally comes from artificial sources and this does not seem like recycling. Of course, it must be so in broad terms. Factories making nitrogenous fertilizer are recycling atmospheric nitrogen in exactly the same way as legumes do, balancing the gaseous losses of nitrogen that occur throughout living systems. Occasionally, a major blockage occurs in recycling systems. Eutrophication of freshwater lakes is an example of this, where undesirable enrichment of nitrogen and phosphates often results in algal "blooms", with eventual loss of nitrogen as methane. The enormous deposits of "guano", mentioned earlier, also represent a failure of the minerals derived from marine fish to be returned to

TABLE 4.5 The nutrients contained in animal excreta

Animal	Composition of the fresh excrement														
	% Nitrogen			% Phosphorus			% Potash			% Lime		% S	% Ca	% Fe	% Mg
	Solid	Liquid	Total	Solid	Liquid	Total	Solid	Liquid	Total	Solid	Liquid				
Horses	0.5	1.2	0.6	0.13	Trace	0.09	0.2	1.2	0.5	0.15	0.45	0.06	0.7	0.01 / 0.1	1 / 3
Cattle (beef)	0.32	0.95	0.5	0.09	0.01	0.87	0.13	0.79	0.5	0.34	0.01				1 / 2
Cattle (dairy)			0.5			0.09			0.4			0.04	0.25	0.04 / 0.1	3
Sheep	0.65	1.68	1.3	0.2	0.01	0.2	0.19	1.7	0.9	0.46	0.16	0.08	0.5	0.01 / 0.2	1 / 3
Pigs	0.6	0.3	0.4 / 0.2	0.2	0.05	0.1 / 0.87	0.37	0.83	0.3 / 0.17	0.09	0	0.1	0.5	0.03 / 0.07	1 / 3 / 2
Hens	1.00	—	1.7	0.35	—	0.61	0.33		0.58						1 / 2

Sources: 1, Martin *et al.* (1976); 2, Gibb and Nielsen (1976); 3, Loehr (1968)

TABLE 4.6 The quantities of excreta produced by agricultural animals

Animal	Solid	Liquid	Total	Source
	\multicolumn{2}{} kg per head per day			
Horses	16.1	3.6	19.7	1
Cattle	23.6	9.1	32.7	1
Cattle			41	2
Dairy cow			36–54	3
Beef cattle			19.8–28	3
Sheep	1.1	0.6	1.7	1
Pigs	2.7	1.6	4.3	1
Pigs			4.5	2
Pigs			4.2–9.3	3
Hens	0.045	—	0.045	1
Hens			0.145	2
Poultry			0.08–0.7	3

Sources: 1, Martin *et al.* (1976); 2, Gibb and Nielsen (1976);
3, Weller and Willetts (1977)

the sea. Agriculture, in this particular case, serves to release the accumulation, ultimately back into the sea.

The biggest drain of nutrients from agricultural systems in the world is represented by human sewage flowing into the sea and, although fish and fish waste-products are used as fertilizer, there is no large-scale return of these nutrients to the land. This might happen in the future, if the processing of sea-water should ever be economical on a massive scale, either to obtain fresh water or deuterium as a source of nuclear power. In the latter case, these would also be a ready source of support energy, although the energy costs of processing large quantities of sea-water would also be very high.

The advantages of recycling *within* systems lie mainly in the savings of energy that result from limitations on transportation and processing costs. This would be true for human sewage also, but since farms feed many more people than the number that live on them, the amount involved would be trivial on large enterprises. Although there are health hazards in using human sewage, it *is* used in many parts of the world (often known as "night soil") and it seems unlikely that we can continue to waste a usable resource on the scale which occurs at present (Table 4.7). However, it could be used for other purposes, particularly as a source of energy, and it may be that the major items for recycling would be phosphate and potash.

This illustrates very well two quite different reasons for recycling. The first is to avoid wastage of trapped solar energy by exploiting all organic

TABLE 4.7 Human Sewage as a resource

Country	Year	Quantity of sewage per head per day	Dry weight of solids from domestic sewage (million t per year)	Nutrients available from domestic sewage (t × 10³)		
				N	P	K
England and Wales	1970	20–60 gallons[a]	1.1[b]	26.9[b]	14.2[b]	3.4[b]
USA[c]	1930	1 141 g[d]	2.6	320	120	100
Hong Kong	1971	1 153 g	0.11[e]	10[e]	5[e]	4.3[e]

[a]Hubbard, 1976
[b]D.O.E. (Welsh Office), 1970
[c]USA data from Snell, 1943
[d]Newcombe, 1977
[e]Assumes same composition as for USA data above

matter until it is completely decomposed. The second is to recover minerals that have to be mined (at some cost) and before they are distributed widely at low concentrations.

It may seem surprising that there has been so little development of crop systems designed to use sewage directly, since it represents the essentials of nutrient and water supply for plant growth. The difficulties and disadvantages, of oxygen supply, for example, are clear but there are plants that might tolerate such conditions and use the environment created with considerable efficiency. There are limits to the amount of sewage that can be absorbed by existing aquatic and terrestrial systems: the question is whether a largely sewage medium can be exploited by any form of crop production.

This is part of an even larger question as to the extent to which protected cropping might be increased in the future.

Protected cropping has more to do with greater control of the environment than of the nutrient supply, as such, but it influences the possibilities of more controlled nutrient supply (as in hydroponics (Cooper, 1975; Cooper, 1976a; Cooper, 1976b; Sholto-Douglas, 1976)) and whether it is worthwhile to achieve this because it reduces losses.

References

Cooke, G. W. (1967). "The Control of Soil Fertility", Crosby Lockwood, London.
Cooke, G. W. (1972). "Fertilizing for Maximum Yield", Crosby Lockwood, London.
Cooper, A. J. (1975). Crop production in recirculating nutrient solution. *Scientia Horticulturae* 3, 251–258.
Cooper, A. J. (1976a). Crop technique: the nutrient film technique. *Horticulture Industry, 1976*, 26–27.
Cooper, A. J. (1976b). The current position of NFT cropping. *Proceedings of the National Glasshouse Conference 1976*, 24–33. Kinsealy Research Centre, Ireland.
DOE (Welsh Office) (1970). Taken for granted. Report of the Working Party on Sewage Disposal. H.M.S.O., London.
Eddowes, M. (1976). "Crop Production in Europe", Oxford University Press, Oxford.
F.A.O. (1975). 1974 Annual Fertilizer Review. F.A.O., Rome.
Geus, Jan G. de (1973). "Fertilizer Guide for the Tropics and Subtropics", Centre d'Etude de l'Azote, Zurich.
Gibb, J. A. C. and Nielsen, V. C. (1976). Farm wastes: animal excreta. *In* "Food Production and Consumption", (Eds A. N. Duckham, J. G. W. Jones and E. H. Roberts), North-Holland Publishing Company, Amsterdam.
Hubbard, C. B. (1976). Sewage sludge — a natural resource or a disposal problem. B.Sc. Honours Dissertation, Department of Soil Science, University of Reading.
Larsen, S. (1973). Recycling of phosphorus in relation to long term soil reserves. *Phosphorus in Agriculture* No. 61, 1–6.
Loehr, R. C. (1968). Pollution implications of animal wastes — a forward-oriented review. U.S. Dept. Interior, Fed. Water Pollution Control Admin, Robert S. Kerr, Water Res. Center, Ada, Oklahoma.

M.A.F.F. (1973). Fertilizer recommendations. Bull. 209. H.M.S.O. London.
Martin, J. H., Leonard, W. H. and Stamp, D. L. (1976). Principles of Field Crop Production. 3rd edn, Macmillan, N. Y. and London.
McCance, R. A. and Widdowson, E. M. (1960). The Composition of Foods. Medical Research Council, Special Report Series No. 297. H.M.S.O., London.
Morrison, J. (1975). Uptake and utilization of nitrogen by forage crops. A. Rep. Grassld Res. Inst. Hurley 1974.
Morrison, J. and Jackson, M. V. (1976). The response of grass to fertiliser nitrogen, the influence of climate and soil. Span, 19(1), 34–37.
Newcombe, K. (1977). Nutrient flow in a major urban settlement: Hong Kong. Human Ecology 5, 179.
Purseglove, J. W. (1972). "Tropical Crops, Monocotyledons, 1", Longmans, London.
Sholto-Douglas, J. (1976). "Advanced Guide to Hydroponics", Pelham Books, London.
Snell, J. R. (1943). Anaerobic digestion. III. Anaerobic digestion of undiluted human excreta. Sewage Works J. 15(4), 679–700.
Söderlund, R. and Svensson, B. H. (1976). The global nitrogen cycle. In "Nitrogen, Phosphorus and Sulphur — Global Cycles", (Eds B. H. Svensson and R. Söderlund), SCOPE Report 7 Ecol. Bull. (Stockholm) 22, 23–73.
Tatchell, J. A. (1976). Crops and fertilizers: overall energy budgets. In Conservation of resources. Proc. Symp. Chem. Soc., Glasgow, 1976.
Weller, J. B. and Willetts, S. L. (1977). "Farm Waste Management", Crosby Lockwood Staples, London.

5

REPRODUCTION IN PLANTS

The main reason for separate consideration of this aspect of plant development is the special significance that method and rate of reproduction may have on efficiency of production. The importance is obvious and direct where reproductive organs or their derivatives represent the products; rather less obvious, but not always of less significance, are the indirect effects of reproduction on such processes as stem and foliar growth.

Methods of Reproduction

Plants reproduce in a great variety of ways (see Table 5.1 for agricultural examples), varying from sexual to vegetative, often adapted to particular environments.

Agriculturally, considerable control may be exercised over the plant population being established and much attention may be devoted to avoidance of losses of reproductive material. Thus the sowing of seeds may be subjected to considerable control, including seed treatment and great care taken in the placement of seed and in the preparation of the environment into which it is introduced. Agricultural practices rarely depend, for example, on seed dispersal by wind or insects.

With pollination, however, the situation is quite different and most pollination of agricultural crops (Ash *et al.*, 1961; Free, 1970; Heinrich, 1975; Start and Marshall, 1976) is carried out by natural agencies (Table 5.2). This example is interesting because the pollination mechanism may not be a very noticeable one, so failure or cause of failure may not be immediately apparent, yet clearly would make an enormous difference to the efficiency of some production systems.

TABLE 5.1 Methods of reproduction in plants

Methods	Reproductive Structures	Examples
1. Sexual		
(a) Bisexual flowers	seeds	most crop plants — wheat, rice, cowpeas, soyabeans, onions
(b) Unisexual flowers	seeds	
monoecious		maize, castorbean, wild rice, cucumber, melon
dioecious		hemp, hops, buffalo grass
(c)	spores	mushrooms
2. Asexual (vegetative)		
(a) Stems	cuttings	sugar cane, Napier grass, cassava, sweet potato
	suckers	bananas
	tubers	potato, Jerusalem artichokes
	rhizomes	Bermuda grass, mint, couch grass
	stolons	strawberry, buffalo grass
	corms	banana, taro
(b) Roots	tubers	sweet potato
(c) Leaves	bulbs or cloves	garlic, shallot
(d) Mycelia		mushroom
(e) Gammae		sponge
(f) Apomictic seed	seeds	Kentucky bluegrass, citrus, mango

TABLE 5.2 Methods of pollination of agricultural crops

Method	Examples of crops primarily pollinated by this method
Self-fertilization	Pea, soyabean, linseed, oats, wheat
Wind (anemophalous)	Maize, sugar beet, most grasses
Insects[a] (entomophalous)	Cotton, cashew, lucerne, clovers, broad bean, hairy vetch

[a]Other animals, especially bats and birds, visit flowers and transfer pollen but do not appear to be important pollinators of agricultural crops.

Agricultural Importance of Reproduction

A wide variety of crop reproductive organs are used for human food or animal feed (Table 5.3) and parts of them are often used as non-food products (Table 5.4).

TABLE 5.3   Plant reproductive organs used for human food or animal feed

Reproductive organs	Examples of food or feed
Tubers (stem and root)	Potatoes Jerusalem artichokes Sweet potatoes
Nuts	Coconuts Hazel nuts Chestnuts Walnuts Brazil nuts Acorns Peanuts
Seeds	Cereals Oilseeds Peas Beans Peppers Nutmegs Vanilla
Fruits	Grapes Raspberries Strawberries Apples Pears Tomatoes Dates
Flowers	Globe artichokes Cloves Cauliflowers Broccoli
Bulbs and corms	Shallots Garlic Cocoyams (taro)
Fruiting bodies of fungi	Mushrooms

TABLE 5.4 Non-food products from plant reproductive organs

Product	Plant	Origin
Fibre		
cotton	*Gossypium* spp.	Seed hairs
coir	*Cocos nucifera*	Husk of fruit
kapok	*Ceiba pentandra*	Seed hairs
Drugs		
opium	*Papaver somniferum*	Unripe capsules
cannabis	*Cannabis sativa*	Dried flowering tops
nux vomica	*Strychnos nux-vomica*	Seeds
Masticatories		
betel	*Areca catechu*	Seed or nut
cola	*Cola nitida*	Seed or nut
Oils (industrial)		
castor	*Ricinus communis*	Seed
linseed	*Linum usitatissium*	Seed
tung	*Aleurites fordii* *Aleurites montana*	Seed
Oils (perfumes)		
rose	*Rosa damascena*	Flower
lavender	*Lavendula officinalis*	Flower
lemon	*Citrus limon*	Peel or fruit
Pectins for pharmaceuticals, cosmetics, etc.	Citrus and apple wastes *Citrus* spp. and *Pyrus malus*	Peel and pulp of fruit
Dyes		
annatto	*Bixa orellana*	Pulp surrounding seed
saffron	*Crocus sativus*	Stigmas
safflower	*Carthamus tinctorius*	Flower head
Insecticide		
pyrethrum	*Chrysanthemum* spp.	Flower
Container — gourd	*Lagenaria siceraria*	Fruit
Decorative flowers	Orchids, carnations, roses, tulips, chrysanthemum, etc.	Flower

Sources: Hill (1952); Schery (1972)

Reproductive processes are the main determinants of production in such cases, but it is not necessarily a simple matter to decide on the most important aspects of reproduction or the best way to describe them. For example, there are interactions between size and number of organs, so it depends upon whether the product has to be of a certain size or whether sheer numbers are important or whether it is the total weight of production that matters.

Where the product is not a direct result of reproduction (e.g. foliage, stems, roots), the quantity and the rate at which it is produced may nevertheless be influenced by reproductive processes. Herbage growth in grasses and legumes is greatly affected by reproduction, most rapid growth occurring during the reproductive period (Spedding and Diekmahns, 1972), whereas root development in beet and radish, for example, may be prevented by too early onset of the reproductive phase.

Crop plants are likely to vary in efficiency, therefore, according to their reproductive state and the level of environmental factors that determine this.

The most important agricultural food crops are established by reproductive means, but some are established vegetatively (Table 5.5). Since all crops have to be newly established at some time, the proportion of plant production that has to be used for this purpose is of some consequence. If the means of establishment is the same as the product (e.g. seed in grain crops, tubers in potato crops), then the quantity needed to sow the next crop has to be subtracted from the gross yield to give a net figure. The proportion to be retained varies with the crop species (Table 5.6), the environmental hazards

TABLE 5.5   Vegetatively-established food crops

Major crops	Other crops	Crops established by a cutting grafted onto a rootstock grown from seed
Potato	Jerusalem artichoke	Apple
Sweet potato	Strawberry	Peach
Cassava	Pineapple	Pear
Sugar cane	Date	Citrus
Banana	Fig	Tung
	Rhubarb	Tea
	Globe artichoke	Coffee
	Shallot	
	Garlic	

Sources include: Edmond *et al.* (1975)

where it has to be established and, above all, with the level of yield: a very low yield might, theoretically, only exceed that needed to establish the next crop by quite a small margin. The proportion needed is also greatly affected by the frequency with which crops have to be replaced: the seed required to replace a tree with a life of 20–30 years is clearly a negligible fraction of the tree's life-time production.

TABLE 5.6 Proportion of crop used to establish another crop

Crop	Yield (kg ha^{-1})	Seed required (kg ha^{-1})	Proportion (%)
Potato (UK)	28 000	2 500	8.9
(Tropics)	10 000	1 500	15.0
Sweet potato[a]	20 000	470	2.4
Wheat (UK)	4 500	190	4.2
Maize (USA)	6 000	up to 20	0.3
(UK)	4 600	25	0.5
Rice (Spain)	5 000	30 (transplanted)	0.6/2.0
(Asia)	1 500	100 (broadcast)	2.0/6.7
Soyabean (USA)	2 242	56	2.5
Sugar cane (Trinidad)	180 000– 300 000	7 500– 10 000	3–4

[a]Most common form of propagation is by vine cuttings

In general, this will be true for perennials as compared with annuals and, when the other advantages of perennials are considered, it is difficult to understand why so much of the world's food comes from annuals. One reason may be that rapid selection is possible because of their generally shorter generation time (from seed to seed).

Perennials and Annuals

The advantages of perennials may be listed as follows:

low *annual* replacement cost;
reduced vulnerability to environmental change (due to well-established root systems, for example);
Maximum opportunity for photosynthesis (leaf area and duration not limited by cultivation methods).

Their disadvantages lie chiefly in the same properties of being present the whole time, and thus able to take advantage of opportunities but continuously exposed to dangers:

exposure to adverse climatic conditions (e.g. extreme cold, heat, water shortage, flooding);
vulnerability to defoliation by the larger animals;
vulnerability to pest and disease attack (both above and below ground).

It may be argued that all these disadvantages apply to annuals when they are growing and to their reproductive organs when they are not. However, one major disadvantage that is characteristic of perennials is the long time that may elapse before the first harvest (e.g. especially tea).

The differences relate to the adaptations of the plant to the two phases of growth and reproduction. Defoliation is particularly serious when temperatures do not allow replacement growth: the seeds of annuals are small and may be protected by hard coats, spines and other structures, as well as lodging in relatively inaccessible places. Agricultural seeds can be stored in complete safety and the plants are mainly exposed to favourable environments.

However, perennials also adapt to seasonal changes in the environment, shed their leaves and reduce their active regions to a small number of well-protected growing points and buds. Furthermore, perennials may produce just as many seeds as annuals do, although this may represent a waste of resources, agriculturally, where the seed is not the product.

Agricultural operations are sometimes easier in the absence of a crop and thus fit better with annuals. This applies to ploughing, in so far as this is concerned with, for example, weed control, but can hardly be seen as an advantage where ploughing is mainly directed at the preparation of a seed bed. One major advantage of annuals is the opportunity provided for crop rotations, as a means of controlling the build up of pests, diseases and weeds, and multiple cropping, especially in the tropics (Stewart, 1970).

At all events, over the world as a whole, a very high proportion of the human diet comes from annual crops and perennial crops provide relatively few of the staple foods (Table 5.7).

Reproductive Efficiency

This may be expressed in several ways. The output may be the units of propagation (seeds, bulbs, tubers) or the established new individual plants. They may be expressed as the number of units or their total weight, or reference may be made to individual unit size (average or range).

TABLE 5.7 Staple foods[a]

Cereals	Rice	Annual
	Wheat	Annual
	Maize	Annual
	Sorghum	Annual
	Barley	Annual
Sugar crops	Sugar cane	Perennial
	Sugar beet	Annual
Root crops	Potato	Annual
	Sweet potato	Perennials
	Cassava	but often grown as annuals
Legumes	Common bean	Annual
	Soyabean	Annual
	Peanut	Annual
Tree crops	Coconut	Perennial
	Banana	Perennial

[a]Source: National Acadamy of Sciences (1975)

The contextual description has to include the environment and may require considerable detail of light intensity and duration. Reproductive rates have to be related to a period of time, but this may be a year or over the life of the plant. The sheer numbers of propagules produced, even in higher plants, is quite astounding (Table 5.8) but they are often individually very

TABLE 5.8 Number of propagules produced by plants

	No. of seeds
Individual plants	
Cabbage	10 000
Tomato	4 000
Sorghum	2 000
Cowpea	850
Crops	No. of seeds per ha
Carrot	615 000 000
Lettuce	361 500 000
Onion	202 700 000
Tomato	28 500 000
Pea	up to 16 400 000

Sources include: Shoemaker (1953)

small (Table 5.9). Because of this, numbers are not always the best measure of output.

TABLE 5.9 Size of propagules (after Martin *et al.*, 1976)

Representative crops	Single seed weight (g)
Redtop, carpetgrass, timothy, bluegrass, fescues, white clover, alsike clover and tobacco	0.00009 −0.0015
Alfalfa, red clover, sweet clover, lespedeza, crimson clover, ryegrass, foxtail millet and turnip	0.0015 −0.003
Flax, sudangrass, crotalaria proso, beet (ball of several seeds), broomcorn, and bromegrass	0.003 –0.009
Wheat, oats, barley, rye, rice, sorghum, buckwheat, hemp, vetch, and mungbean	0.009 –0.045
Corn, pea and cotton	0.045 –0.88
Potato and Jerusaem artichoke	22.7–113

Whatever the measure chosen, it is usually expressed per plant or unit area of land, or more rarely, per unit weight of plant. A great many factors usually influence reproductive efficiency and the manipulation of these is a major part of the agricultural activity, since grain crops provide 53% of man's food energy (Brown and Finsterbusch, 1972).

References

Ash, J. S., Hope Jones, P. and Melville, R. (1961). The contamination of birds with pollen and other substances. *British Birds* **54**, 93–100.
Brown, L. R. and Finsterbusch, G. W. (1972). "Man and His Environment: Food", Harper and Row, New York.
Edmond, J. B., Senn, T. L., Andrews, F. S. and Halfacre, R. G. (1975). "Fundamentals of Horticulture", 4th edn, McGraw-Hill, New York.

Free, J. B. (1970). "Insect Pollination of Crops", Academic Press, London and New York.

Heinrich, B. (1975). Energetics of pollination. *In* "Annual review of Ecology and Systematics", (Eds R. F. Johnston, P. W. Frank and C. D. Michener), Vol. 6, 139–170. Annual Reviews Inc., Palo Alto, CA.

Hill, A. F. (1952). "Economic Botany", 2nd edn, McGraw-Hill, New York.

Martin, J. H., Leonard, W. H. and Stamp, D. L. (1976). "Principles of Field Crop Production", 3rd edn, Macmillan Publishing Co., New York.

National Academy of Sciences (1975). Agricultural Production Efficiency. N.A.S., Washington.

Schery, R. W. (1972). "Plants for Man", 2nd edn, Prentice-Hall, Englewood Cliffs, New Jersey.

Shoemaker, J. S. (1953). "Vegetable Growing", 2nd edn, Wiley, New York.

Spedding, C. R. W. and Diekmahns, E. C. (Eds) (1972). "Grasses and Legumes in British Agriculture", Commonwealth Agricultural Bureaux, Farnham Royal, Buckinghamshire.

Start, A. N. and Marshall, A. G. (1976). Nectarivorous bats as pollinators of trees in West Malaysia. *In* "Tropical Trees", (Eds J. Burley and B. T. Styles). Academic Press, London and New York.

Stewart, G. A. (1970). High potential productivity of the tropics for cereal crops, grass forage crops, and beef. *J. Aust. Inst. agric. Sci.* **36**, 85–101.

6

GRAIN PRODUCTION

The grain crops are of the greatest importance in feeding the world's population. Quite a wide range of species is used but most of them rather locally in small quantities. The major species are given in Table 6.1, with their main cultural requirements.

Strictly speaking, a grain is a caryopsis, a dry indehiscent fruit containing one seed, and may be defined as the fruit of cereals. Most of the cereals (and *all* by some definitions) are grasses, i.e. members of the family Gramineae. This is true of all the major cereals of the world — wheat, rye, barley, maize, oats and rice, and of many of the less important ones, such as millet, but is not true, for example, of buckwheat.

Legumes are also grown for their seed and these are sometimes referred to as grain legumes: here, we are calling them "pulses" and dealing with them in a separate chapter (Chapter 7).

Grain crops are grown almost entirely for their seeds, either for direct human consumption or for feeding to animals, but there is increasing interest in using the straw as a by-product and it should not be forgotten that their root-systems also represent a useful source of organic matter added to the soil.

However, in terms of efficiency of production, there is little difficulty in defining the product. Furthermore, it is usually relatively dry at the point of storage, though not necessarily so as harvested (see Table 6.2 for the normal variation in moisture content), and can thus be compared fairly easily between species. Grains are usually thought of as primarily a source of dietary energy but, in view of the extent to which they contribute to human diets, their protein output cannot be ignored. It has been suggested (Thielebein, 1969) that cereals provide about half the total human requirement for protein and Whitehouse (1973) has calculated that the yield data for 1968, taking average values for protein content, represented enough protein to supply 100 g per head per day to a population of 3000 m people. This

74

TABLE 6.1 The main grain crops

	Species	Main cultural features
Wheat	*Triticum aestivum*	Cool moist growing season, warm dry harvest. Fertile, well-drained soil.
Rice (lowland)	*Oryza sativa*	Irrigation water, mean temperature of 70°F or above for four- to six-month growing season.
Maize	*Zea mays*	Warmth and adequate moisture, fertile, well drained soil. Optimum pH 6.0–7.0. High nitrogen requirement
Barley	*Hordeum vulgare*	Cool growing season. Well drained soil pH 6 or above. Most dependable cereal under extreme conditions of salinity, summer frost or drought.
Oats	*Avena sativa*	Cool, moist climate. Does not require as high fertility as most cereals.
Rye	*Secale cereale*	Hardiest, and more productive on in-fertile, sandy or acid soils than other cereals.
Sorghum	*Sorghum vulgare* or *bicolor*	Can withstand periods of drought and extreme heat; will tolerate consider-able soil salinity. Killed by frost. Will tolerate temporary waterlogging.
Millet		
Pearl or Bulrush millet	*Pennisetum typhoides* or *glaucum*	Most productive grain in extremely dry or infertile soils of India and Africa. Will not tolerate water-logging. Susceptible to bird damage.
Foxtail millet	*Seteria italica*	Similar to Pearl millet
Proso or Common millet	*Panicum miliaceum*	
Finger millet	*Eleusine caracana*	Requires more moisture than the other millets.

illustrates one major problem in any comparison of proteins, since grain crops vary in the proportions of amino acids that are present (Table 6.3). This situation can be represented by calculations of biological value (B.V.) but only total crude protein (N × 6.25) will be used here to illustrate the efficiency of protein production in cereals. The effect of the low biological value is that at least 50% more of such protein must be ingested by non-ruminant animals than would otherwise be necessary.

TABLE 6.2   Moisture content of grains

	At harvest (%)	Required for storage (%)
Wheat	16–24	14–15
Rice	18–27	12–15
Maize	15–35	12–15
Barley	—	15
Oats	—	11
Rye	—	10
Sorghum	15–20	8–16
Millet	—	12–16

The most important resources for cereal production, in general, are land (including its function as a receiving surface for solar radiation) and power.

Of course, cereals cannot be grown without oxygen, carbon, water, and all the other mineral requirements for plant growth, and there has to be a supply of seed and this may require money. Given an agricultural environment, however, the key inputs for efficiency calculations are land and power. The amount produced per unit of land (per unit time, most commonly per year) is the most important ratio: it is influenced by other inputs but the only one that is vital, over and above those contained in the environment, is a source of power. The crop has to be sown in prepared soil, it has to be weeded and harvested, and all these activities require power. The latter can be supplied by human labour, animals, or machinery powered by wind, water or "support" energy from fossil fuels. The need for human labour can only be diminished by use of another power source and this always has a cost. In the case of animals, the cost may be the diversion of food grown to feed the animals: in the case of fossil fuels, there is a financial cost that is likely to increase as the supply of such fuels runs out or the cost of extracting new reserves increases. Wind and water power are methods of indirect harnessing of more solar radiation but they are not always readily

TABLE 6.3 Amino acid content of cereal grains (g per 16 g nitrogen)

	Wheat	Rice	Maize	Barley	Oats	Rye	Sorghum
Arginine	4.3–4.7	7.7–7.8	4.3–5.0	5.0–5.4	6.1–6.6	5.0	2.7–4.7
Cystine	2.1	1.1	2.1	2.1	1.8	1.8	1.0
Histidine	2.1–2.3	2.2–2.3	2.4–2.6	1.9–2.2	1.9–2.2	2.1	2.0–3.3
Isoleucine	3.4–3.8	3.9–4.5	3.8–4.0	3.7–3.8	4.0–4.6	3.9	3.8–4.7
Leucine	6.4–6.8	8.0	10.6–12.0	6.9–7.1	7.0–7.1	6.1	11.6–14.3
Lysine	2.6–2.7	3.5–3.7	2.7–3.0	3.4–3.7	3.7–4.0	3.7	1.8–2.9
Methionine	1.6	2.4	2.1	1.4	1.6	1.6	1.6
Phenylalanine	4.6	5.2	4.6	5.0	5.0	4.6	4.3
Threonine	2.9–3.0	3.3–4.1	4.0–4.2	3.6–3.7	3.4–3.6	3.6	3.0–3.8
Trytophan	1.1–1.3	0.6–1.4	0.7–0.8	1.3–1.4	0.9–1.3	1.3	0.7–0.8
Tyrosine	3.2	3.3	3.8	3.5	3.8	4.2	2.7
Valine	4.3–4.6	5.4–5.7	5.0–5.6	5.0–5.3	5.1–5.4	5.0	4.9–6.0

Sources: Kent (1966); Whitehouse (1973)

TABLE 6.4 Output of energy and protein per unit of land

Grain crop	Range of DM yield per ha per yr (kg)	Record yield (kg)	Energy content (MJ kg⁻¹ DM)	Protein content (%)	Energy output (GJ ha⁻¹)	Protein output (kg ha⁻¹)
Wheat	2 000–5 000	14 528	18.4	8–20	37–92 267[a]	160–1 000 2 905[a]
Rice	2 000–6 000		18.0	6–15	36–108	120–900
Maize	3 000–6 000	19 272	19.0	8–15	57–114 366[a]	240–900 2 890[a]
Barley	2 000–4 000	11 407	18.3	9–27	32–73 209[a]	180–1 080 3 080[a]
Oats	1 000–5 000	10 618	19.0	11–26	19–95 201[a]	110–1 300 2 760[a]
Rye	2 000–4 000		18.4	8–13	37–74	160–520
Millet	1 000–3 000		18.7	7–12	19–56	70–360
Sorghum	1 000–5 000		18.8	8–20	19–94	80–1 000

[a]Record yields
Sources: F.A.O. (1975); Wittwer (1975); M.A.F.F. (1975)

TABLE 6.5 The output of energy and protein per unit of labour

| | | Yield per hour of labour | | |
Crop	Country	Grain (kg)	Energy[a] (MJ)	Protein[b] (kg)
Wheat	UK	278	5 115	34.5
	USA	300	5 520	37.2
Barley	UK	252	4 612	27.2
Maize	USA	417	7 923	40.9
Sorghum	USA	386	7 257	41.7
Rice	Nigeria	1.3	23	0.1

[a]Energy contents as for Table 6.4
[b]Protein contents: M.A.F.F. (1975) Tables of Feed Composition and Energy Allowances for Ruminants

Sources: Nix (1976); U.S.D.A. (1973); Phillips (1964)

TABLE 6.6 The output of energy (GJ) and protein (kg) per unit (GJ) of support energy

Crop	Country	Source	Energy	Protein
Rice	USA	Leach (1975)	1.3	8
Rice	Thailand	Poonsab (1976)	3.7	16
Maize	USA	Pimental (1973)	2.8	14
Barley	UK	Leach (1975)	3.6	21
Oats	UK	Leach (1975)	3.5	20
Wheat	UK	Leach (1975)	4.2	28
Sorghum	USA	Heichel (1973)	5.8	33

TABLE 6.7 The output of grain per unit of applied fertilizer nitrogen

Crop	Country	N input (kgha^{-1})	Grain yield (kg ha^{-1})	Output per kg N (kg)
Wheat	UK	97	4500	46
Rice				
Indica cvs	SE Asia	40	1500	38
Japonica cvs	Spain Japan	150	5000	33
Maize	USA	30–160	6270	39–209
	UK	100	4600	46
Barley	UK	110	3900	35
Oats	UK	94	4250	45
Sorghum	USA	180	6500	36
	India	45	1800	40

Sources: Nix (1976); Purseglove (1972); Martin *et al*. (1976)

TABLE 6.8 An analysis of support energy inputs for rice production (using a 65 h.p. tractor and recommended fertilizer levels) in Thailand (after Poonsab, 1976)

Input	MJ	% of total support energy
Rice seed as seedlings for planting	288.9	2.6
Diesel to power a tractor	2 166.3	19.3
Petrol to power a pumping machine	2 901.1	25.8
Petrol to power a pesticide spraying machine	124.0	1.1
Petrol for threshing and winnowing	3 793.3	33.7
Pesticide	39.9	0.1
Tractor (depreciation, repair, oil and greases)	587.1	5.2
Nitrogen in fertilizer	999.3	8.9
Phosphate in fertilizer	349.8	3.1

TABLE 6.9   Efficiency of use of solar radiation with different levels of inputs and outputs

N input (kg ha^{-1})	Wheat ^{a} grain yield (kg ha^{-1})	$E = \dfrac{\text{Grain yield}}{\text{Solar rad}^b} \times 10^{-3}$
0	2 637	0.08
	3 390	0.10
50	3 767	0.11
	4 196	0.127
100	4 268	0.129
	6 278	0.19
150	4 394	0.133
	6 529	0.198

^{a}Two varieties of wheat and under different conditions
^{b}Total annual radiation receipt of 33 × 10^6 MJ ha^{-1}yr^{-1}

Source: Whitear (1972)

TABLE 6.10   Water transpired per unit of grain produced

Crop	kg water per kg grain produced^a
Rice	1 420
Rye	1 370
Oats	1 194
Maize	1 104
Barley	1 068
Wheat	1 026
Sorghum	966
Millet	930

^{a}Water requirements per gram of dry matter produced from Hildreth et al. (1941).

Requirements for grain production calculated assuming grain:straw ratio of 1:1 for wheat, rice, barley, oats and rye, 1:2 for maize, sorghum and millet.

TABLE 6.11 Optimum[a] day
and night temperatures for
some cereals (from Friend,
1966).

| | Temperature (°C) | |
	Day	Night
Wheat	25	25
Oats	20–30	25–30
Maize	30–35	25

[a]Optimum vegetative growth

available where they are wanted and the mechanisms and equipment required to harness them also have costs, both in financial and support energy terms.

The contribution that can be made by human labour can be much modified by the application of human skills and, indeed, one way of viewing the input of power to agriculture is in terms of an increasing input of skill rather than energy. The first step is to think about how human energy is deployed directly, the next to increase the effect of applied energy by the use of tools, and, finally, to substitute fuel for the human energy used to power the tools. The significance of human "labour" in its widest sense is not, therefore, well expressed energetically. It seems more useful to consider labour and support energy separately, expressing output per unit of each resource directly.

The major measures of efficiency are thus output of energy and protein per unit of land, labour and support energy. These are illustrated for "normal" crops, for regions where they are widely grown, in "normal" years, with accepted levels of fertilizer input, in Tables 6.4–6.6.

Since the normal input of fertilizers differs for different crops, however, the output per unit of applied fertilizer also varies (see Table 6.7).

An analysis of the support energy inputs (Table 6.8) shows where economies might conceivably be made but it has to be remembered that any change that results in a lowered yield per unit of land has the effect of reducing the efficiency with which solar radiation is used (Table 6.9).

Yield may also be greatly influenced by water supply and cereal crops do differ somewhat in their water requirements (Russell, 1966; Slavik, 1966) and in the quantities of water normally transpired per unit of grain produced (Table 6.10).

There are also differences between cereals in response to light and temperature (Friend, 1966) and between their optimum day and night temperatures (Table 6.11; Peters et al., 1971).

Factors Affecting Yield

Most of the carbohydrate in cereal grains is probably formed from CO_2 assimilated after the ears emerge (Thorne, 1966). Growth before this time must influence yield indirectly, by affecting the surface available for photosynthesis after ear-emergence and by affecting the number and potential size of sites at which carbohydrate can be stored. The post-ear-emergence photosynthesis may occur chiefly in the leaves (e.g. maize) or in the ears (e.g. barley): the size of this photosynthetic area will clearly influence yield per plant. Yield per unit of land will be affected by the number of ears per ha, the

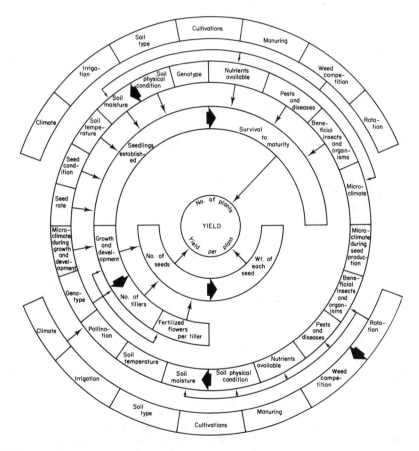

FIG. 6.1 The main factors affecting yields of cereals. Only some of the connections between the circles are shown in the interests of simplicity, i.e. some factors, especially on the outer circles, may affect many of the factors on the inner circles. The diagram can be extended as required.

number of grains per ear and grain size. Very often, however, an increase in one of these may be accompanied by a decrease in others. The main determinant of yield is often different for different cereals and even differs between varieties, especially, for example, spring and autumn varieties. Optimum planting densities differ greatly between cereals (Evans and Wardlaw, 1976).

Where numbers are concerned, it is the final number that matters and this may not be a simple reflection of the number produced initially. The proportion of tillers that survive tends to increase where fewer tillers are produced and the number of ears is influenced by the number of *surviving* tillers.

However, the advantages and disadvantages of *numbers* varies with the incidence of pests and disease, since these also affect the proportions surviving.

The main factors affecting yield therefore vary with environment but they operate through the main components of yield. These are illustrated diagrammatically in Fig. 6.1.

Donald and Hamblin (1976) distinguish between "biological yield" (the total yield of plant material, which should really include the roots but, since these are normally non-recoverable, is usually applied to the total weight of tops); total yield of grain; yield of grain and straw and yield of dry matter. These, when multiplied by the "migration coefficient", the ratio of grain to total production or the coefficient of effectiveness or the "harvest index", give the grain yield. (The "harvest index" is the ratio of grain yield to the biological yield.)

References

Donald, C. M. and Hamblin, J. (1976). The biological yield and harvest index of cereals as agronomic and plant breeding criteria. *Adv. Agron.* **28**, 361–405.

Evans, L. T. and Wardlaw, I. F. (1976). Aspects of the comparative physiology of grain yields in cereals. *Adv. Agron.* **28**, 301–359.

F.A.O. (1975). Production Yearbook 1974. Vol. 28. F.A.O. Rome.

Friend, D. J. C. (1966). The effects of light and temperature on the growth of cereals. *In* "The Growth of Cereals and Grasses", (Eds F. L. Milthorpe and J. D. Ivens), Proc. 12th Easter School in Agricultural Science, University of Nottingham, 1965. Butterworths, London.

Heichel, G. H. (1973). Comparative efficiency of energy use in crop production. Bull. 739. Connecticut Agricultural Experimental Station, New Haven.

Hildreth, A. C., Maguess, J. R. and Mitchell, J. W. (1941). Effects of climatic factors on growing plants. *In* "Yearbook of Agriculture, 1941", U.S.D.A. Washington D.C.

Kent, N. L. (1966). "Technology of Cereals", Pergamon, Oxford.

Leach, G. (1975). Energy and food production. International Institute for Environment and Development, London.

M.A.F.F. (1975). Tables of Feed Composition and Energy Allowances for Ruminants. H.M.S.O., London.

Martin, J. H., Leonard, W. H. and Stamp, D. L. (1976). Principles of Field Crop Production. 3rd edn, Macmillan Publishing Co., New York.

Nix, J. (1976). "Farm Management Pocketbook", 7th edn, Wye College, Ashford.

Peters, D. B., Pendleton, J. W., Hageman, R. H. and Brown, C. M. (1971). Effect of night air temperature on grain yield of corn, wheat and soybeans. *Agronomy J.* **63**, 809.

Phillips, T. A. (1964). "An Agricultural Notebook", Longmans, Nigeria.

Pimental, D., Hurd, L. E., Bellotti, A. C., Forster, M. J., Oka, I. N., Sholes, O. D. and Whitman, R. J. (1973). Food production and the energy crisis. *Science* (N.Y.) **182**, 443–449.

Poonsab, S. (1976). Energy equation of rice production. M.Sc. Thesis. Faculty of Graduate Studies, Mahidol University, Thailand.

Purseglove, J. W. (1972). "Tropical Crops. Monocotyledons 1", Longmans, London.

Russell, E..W. (1966). The soil environment of graminaceous crops. *In* "The Growth of Cereals and Grasses", (Eds F. L. Milthorpe and J. D. Ivens), Proc. 12th Easter School in Agricultural Science, University of Nottingham, 1965. Butterworths, London.

Slavik, B. (1966). Response of grasses and cereals to water. *In* "The Growth of Cereals and Grasses", (Eds F. L. Milthorpe and J. D. Ivens), Proc. 12th Easter School in Agricultural Science, University of Nottingham, 1965. Butterworths, London.

Thielebein, M. (1969). The world's protein situation and crop improvement. *In* Joint F.A.O/I.A.E.A. Division of Atomic Energy in Food and Agriculture. New Approaches to Breeding for Improved Plant Protein. Panel Proceedings Series — STI/PUB/212, 3–6. Vienna, International Atomic Energy Agency.

Thorne, G. N. (1966). Physiological aspects of grain yield in cereals. *In* "The Growth of Cereals and Grasses", (Eds F. L. Milthorpe and J. D. Ivens), Proc. 12th Easter School in Agricultural Science, University of Nottingham, 1965. Butterworths, London.

U.S.D.A. (1973). Agricultural Statistics. Washington, D.C.

Whitear, J. D. (1972). Nitrogen requirements of winter wheat varieties. Fisons Agricultural Technical Information. *AGTEC* Autumn 1972.

Whitehouse, R. N. H. (1973). The potential of cereal grain crops for protein production. *In* "The Biological Efficiency of Protein Production", (Ed. J. G. W. Jones), Cambridge University Press, London.

Wittwer, S. H. (1975). Food Production: Technology and the Resource Base. *In* "Food: Politics, Economics, Nutrition and Research", (Ed. P. H. Abelson), American Association for the Advancement of Science, Washington D.C.

7

PULSE CROPS

The grain legumes are frequently thought of as protein crops because they generally produce seeds with a very high protein content (Table 7.1). There is considerable variation between species, however, and their energy content is also considerable (Table 7.2). The result is that pulse crops are very productive per unit of land (Table 7.3) and, having the enormous advantage that they fix their own nitrogen, they require relatively low inputs of fertilizer. The fact that they need so little nitrogenous fertilizer gives them advantages in efficiency of support energy use (Table 7.4).

The contribution of grain legumes to world food production is important (Table 7.5) but none of them contribute anything like the quantities associated with the main cereals. They are substantial contributors to the protein component of animal feedstuffs (Scott, 1974) but, again, in total they are not as important as the cereals. Considerable research is now being directed to the breeding of legumes and other crop plants with higher protein contents and a better balance of essential amino acids (I.C.R.I.S.A.T., 1975; Protein Advisory Group, 1975).

Grain legumes grow in a wide range of environments (see Table 7.6 for an illustration of optimum temperatures) but tend to require a good water supply and a high labour input (Table 7.7).

Several legume species are climbers and need support for full growth. This makes them suitable for certain mixtures of species, but this has not been very widely exploited.

The moisture content of the seed is generally lower than that of cereals, and several species have a high content of oil (Table 7.8). Two of the main sources of vegetable oil are legumes: they are soyabean and peanuts (Ames, 1975). The latter have a curious growth habit; they start as erect plants but, after the flowers have withered, the stems bearing the pods bend over and force them underground, where they ripen. (Subterranean clover also buries its seeds in this way.) Soyabeans are now grown on a vast scale and are a major component of the commercial trade in animal feedstuffs.

TABLE 7.1 Protein content of grain legume seeds

		Protein (%)
Soyabean	*Glycine max*	25–52
Peanut or groundnut	*Arachis hypogaea*	24–34
Broad bean	*Vicia faba*	24–35
Cowpea	*Vigna unguiculata*	21–24
Lentil	*Lens esculenta*	24–30
Common bean	*Phaseolus vulgaris*	20–25
Pea	*Pisum sativum*	20–30
Chickpea	*Cicer arietinum*	17–26

Sources: Smartt (1976); F.A.O. (1954); M.A.F.F. (1975); Purse-glove (1968)

TABLE 7.2 Energy content of grain legume seeds

	Energy content (MJ per kg DM)
Soyabean	15.2–23.1
Peanut	24.1–29.7
Broad bean	16.0–18.0
Cowpea	15.9
Lentil	16.1–19.1
Common bean	15.9
Pea	16.1–18.9
Chickpea	16.7–18.4

Sources: F.A.O. (1954); M.A.F.F. (1975)

TABLE 7.3   Production of protein and energy by grain legumes

Crop	Grain yield (kg ha^{-1})	Protein yield[a] (kg ha^{-1})	Energy yield[b] (GJ ha^{-1})
Soyabean	1 000–2 700	250–1 404	15.2–62.4
Peanut	500–2 300	125– 782	12.0–68.3
Broad bean	1 000–3 200	240–1 120	16.0–60.8
Cowpea	400–2 800	84– 672	6.4–44.5
Lentil	400–1 700	96– 510	6.4–32.5
Common bean	500–2 000	100– 500	8.0–31.8
Pea	2 000–3 500	400–1 050	32.2–66.2
Chickpea	450–1 800	77– 468	7.5–33.1

[a]see Table 7.1
[b]see Table 7.2

Sources: Smartt (1976); Purseglove (1968); Nix (1976)

TABLE 7.4   The output of energy and protein per MJ of support energy from grain legumes

	Energy (MJ)	Protein (kg)
Peanuts[a]	0.19	0.002
Soyabeans[a]	0.34	0.009
Broad beans[b]	0.78	0.019
Dwarf beans[b]	0.21	0.004
Garden peas[b]	1.27	0.028

[a]Calculated from Heichel (1973)
[b]Shiels (1978). Values are calculated for these crops harvested as green vegetables. Harvesting as pulse crops would probably have little effect on the energy ratio.

TABLE 7.5   Contribution (%) of world food production by pulse crops

	Protein	Energy
Pulses	12.9	5.2
Cereals	52	56

Sources: F.A.O. (1975); Evans (1975); Brown and Eckholm (1975)

TABLE 7.6   Optimum temperatures for pulse crops

Crop	Stage of development or process	Temperature (°C)
Peanut	Germination	15–45 range
		32–34 opt
	Growing	24–33 opt
Cowpea	Maximum dry matter production	27 day
		22 night
Mung bean (*Vigna radiata*)	Germination	24–32
Garden pea (*Pisum sativum*)	Germination	4–30 range
		18 opt
	Dry matter accumulation	21–23 day
	up to 6th node	10–16 night
	beyond 6th node	16–18 day
		10–13 night

Sources: Rachie and Roberts (1974); Sutcliffe and Pate (1977)

TABLE 7.7 Water and labour requirements for pulse crops

Crop	Water requirement[a]	Labour requirement (Man-hours ha^{-1})	Yield ha^{-1b}
Soyabeans	744	11.4	1 836(1)
		15	1 882(2)
		10.6	2 673(3)
Cowpeas	571		
Chickpeas	663		
Beans	728	15.8	2 600(4)
		14.9	2 900(4)
Peas	788	14.8	2 750(4)

[a]kg water transpired per kg plant dry matter produced, from Hildreth *et al* (1941).
[b]Sources for labour requirement and yields: (1) U.S.D.A. (1973); (2) Pimental *et al*. (1975); (3) Caldwell (1973); (4) Nix (1976)

TABLE 7.8 Oil[a] content of legume seeds

	Oil (%)[b]	
Peanut	47.7	(1)
	38–50	(2)
Soyabean	13.5–24.2	(1)
Chickpea	5.3	(1)
Winged bean (*Psophocarpus tetragonolobus*)	15.0	(1)
Ground bean (*Voandzeia subterranea*)	4.5–6.5	(1)

[a]Oil or fat
[b]Sources: (1) Smartt (1976); (2) Purseglove (1968)

Since the seeds are often large or carried in substantial pods, harvesting is fairly easily mechanized, but storage is more difficult for those with high moisture contents, unless they are frozen or dried.

Very often the foliage of pulse crops is also valued as a forage, often ensiled (as with pea-haulm silage), or as a green manure. The foliage also tends to have a high nitrogen content, with the same advantage of requiring negligible inputs of nitrogenous fertilizer.

This feature is of immense importance and is likely to become even more so than in the immediate past. The ability to fix atmospheric nitrogen depends, of course, on the presence in the soil of the appropriate varieties of nodule-forming bacteria. These cannot be assumed to be present wherever they are required and many legume seeds are inoculated before sowing. Where they are not, growth of seedlings may be erratic and levels of production inconsistent. Furthermore, the relevant bacteria may not thrive in all soil types and at particular soil pH values and are also subject to temperature constraints. Clearly, the efficiency of production by legumes depends enormously on such factors, unless nitrogenous fertilizer is supplied. Legumes grow perfectly well with applied nitrogen but rarely grow any better than when effectively nodulated.

Use of nitrogen by legumes well illustrates some important features of efficiency calculations. Superficially, legumes can be said to be vastly more efficient than cereals in their use of applied nitrogen for the production of energy or protein (Table 7.9). Where high levels of nitrogenous fertilizer are actually applied, however, this is not strictly true. Where none at all is applied to either crop, the calculation has no meaning, since there is no input. If a little is applied to the legume, it appears highly efficient, whatever the level applied to the cereal. But, of course, the nitrogen in the crop has to come from somewhere and the significance of the quantities applied depends upon the availability of other sources. These are the soil and the atmosphere and where legumes are effectively nodulated they require no additional, applied nitrogen. This is no different from the situation for a cereal crop immediately following a heavily fertilized grass ley, where the soil has quite sufficient nitrogen for the crop and, equally, requires none applied.

There is thus a useful difference to be recognized between not needing a resource at all and the efficiency with which it is used. Otherwise a crop such as rice, grown in water and needing none added by irrigation, could be said to use irrigation water with an extremely high efficiency. The fact is that the crop *would* respond to water if the latter was in short supply but it would not normally be grown in these circumstances.

It is the great advantage of legumes that they are generally able to make use of a resource — atmospheric nitrogen — that is present in unlimited quantities and freely available in all environments. However, it should be recognized that legumes incur an energy cost in the process of N fixation. A similar situation exists for the supply of oxygen to most terrestrial plants, although there may be problems in the oxygen supply to plant roots in some soils, including water-logged ones, just as there are for aquatic plants. To a

TABLE 7.9 Efficiency of fertilizer nitrogen use by pulse crops

Crop	N use[a] (kg ha^{-1})	Grain[b] yield (kg ha^{-1})	Protein yield[c] (kg ha^{-1})	Protein yield (kg) per kg fert. N used	Energy yield[c] (GJ ha^{-1})	Energy yield (GJ) per kg fert. N used
Peas (UK)	25	2 750	550–825	22–33	44–52	1.8–2.1
Field beans (UK)	25	2 600	624–910	25–36	42–47	1.7–1.9
Winter wheat (UK)	75	4 500	360–900	4.8–12	83	1.1
Soyabeans (USA)	22	1 836	459–955	21–43	28–42	1.3–1.9
Peanuts (USA)	22	2 360	566–802	26–36	57–70	2.6–3.2
Maize (USA)	170	5 800	464–870	2.7–5.1	110	0.6

[a] In most cases N fertilizer is not used at all for pulse crops and the amounts suggested are for very poor soils to support the plant until nodulation is effective

[b] Grain yields used are those given in the reference for fertilizer N usage wherever possible

[c] Protein and energy contents are used in Table 7.1 and Table 7.2

Sources: Martin et al. (1976); Nix (1976)

considerable extent, this is also true for CO_2, but this does not mean that there is no point in calculating the efficiency with which additional CO_2 is used in certain glasshouse situations, for example. This illustrates another facet of efficiency calculations. It was stressed in earlier chapters that no single assessment of efficiency could necessarily demonstrate the value of a given input, and the addition of CO_2 is a clear example. The costs of supplying CO_2 are much more than the cost of the gas itself, since considerable investment in equipment is required to deliver it and it can only be done sensibly within certain controlled environments. The actual efficiency with which the gas is used may therefore be of far less significance than the yield response to its supply. Indeed, it is quite conceivable that one would wish to apply quantities well above the level at which it was used with maximum efficiency, provided that this resulted in greater yields per unit of land or greenhouse. This is often true of many inputs, including fertilizers, and shows how often interest really centres on some overall process, to which a particular constituent process contributes. This is so whether the latter is simply a small part of the total or whether it interacts with other processes. Interactions are common: the effect of irrigation on the supply of soil nitrogen to plant roots is an example.

Very often, then, interest in an input, such as CO_2, is primarily concerned with its effect on the major process and its ability to influence the efficiency of that process. The addition of CO_2 can thus be visualized as an influence on the process of production per unit area of soil or on the output of cash per unit of money invested, or on any of the other major efficiency ratios.

The subject of CO_2 supplementation has been used here to illustrate aspects of efficiency related to gaseous exchanges, some of which are especially relevant to legumes. As it happens, however, CO_2 is rarely supplied to legumes and is more commonly used with glasshouse crops such as tomatoes.

As with cereals, the efficiency of grain legumes is also greatly influenced by the partition of nutrients to the grain and the factors affecting this. Obviously, the number of pods, the number of seeds per pod and the size of each seed must govern grain yield, but the factors affecting each of these and the interactions between them are complex. The number of flowers is an important feature but the proportion pollinated is more important in legumes than many other crops, simply because pollination is rather less certain, often depending (as it does in many legumes) on the activities of insects. Furthermore, in the case of many legumes, only certain groups of insects are capable of pollination. (In other groups of plants, pollination may depend upon different groups of insects, on bats and on birds (see Heinrich, 1975; McGregor, 1976).) This is so for the field bean (*Vicia faba*), for example, which can be most efficiently pollinated by the long-tongued bumble bees (*Bombus agrorum, B. hortorum, B. lapidarius* and *B. ruderatus*). This means that the process is influenced by all those features of the environment (such as temperature and wind-speed) that affect insect activity. Since insects are also subject to disease, predators and parasites, as well as to the availability of food supplies, discrepancies can easily arise between the size of the crop population and the size of the insect population required to pollinate it. This problem can occur in an acute form where a very large area is devoted to one kind of crop, in which pollination has to occur over a relatively short period. The insect population needed must also be large, to accomplish the task in the time available, and has to find other foods for the rest of its active season. For the other foods to be readily available the insect population has to be sited peripherally to the legume crop: the logistical problems could be insuperable, for example, if a beehive was sited in the centre of a vast monoculture.

It is worth noting, in passing, that such factors also operate quite differently in small, experimental areas and in large, practical applications of experimental results.

Once flowers are pollinated, it is chiefly a matter of nutrient supply that determines seed size. The photosynthetic surface available is clearly important in this but it is not possible to generalize further about which part of the plant matters most. In some grain legumes the leaves mainly supply the developing seed but in plants such as the leafless pea (Snoad and Davis, 1972; Snoad, 1974) this is not so and the burden of photosynthesis rests on the stems and tendrils and on the pods themselves.

The factors chiefly concerned in determining seed yield are illustrated in Fig. 7.1.

Increasingly, however, crops may have to be assessed in terms of their total yield, with the foliage, at least, being used as a by-product. This could be for animal feed, by grazing, ensilage or drying, or for a leaf-protein extraction process (Byers and Sturrock, 1965; Oelshlegal et al., 1969), which might produce food for direct human consumption. The leafless pea was produced in order to minimize harvesting problems and, presumably, if it produces less total dry matter, it can be grown at greater population densities.

Another wider issue, that may have to be considered increasingly, is the possibility of greater efficiency of resource use (especially of land, light and nitrogen) being obtained by mixed cropping. Some advantages have been shown in some circumstances (Willey and Roberts, 1976; Rao and Willey, 1978) and advantages of mixing legumes and non-legumes could arise from the production of a better balanced feed for animals, in terms of nitrogen and energy contents, particularly, and perhaps a transfer of some nitrogen, fixed by the legume, to the non-legume (for example, by shedding of nodules). Some of the results of mixed cropping experiments are given in Table 7.10 and also in Table 3.2.

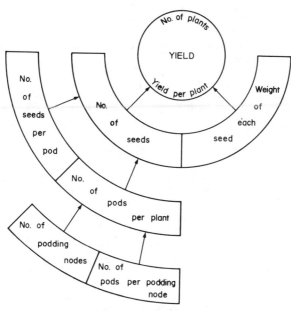

FIG 7.1   The components of yield normally listed for pulse crops (see Hardwick and Milbourne, 1967; Ishag, 1973; Thompson and Taylor, 1977). It is clear that this could be systematically expanded to show all the important factors determining those shown, these would be very similar to the factors shown in Fig. 6.1 for cereals. The important differences being in the method of pollination (mainly by insects for pulse crops instead of wind as for cereals) and the ability of legumes to fix nitrogen.

TABLE 7.10   Some results of mixed cropping experiments

Source	Crop	Country	Yield from a pure stand (kg ha^{-1})	Yield under mixed cropping (kg ha^{-1})
I.C.R.I.S.A.T. (1975)	Chickpea	India	2 466	1 237
	Brassica campestris		1 549	1 018
Rao and Willey (1978)	Sorghum	India	2 027	1 504
	Pigeon pea		1 016	728
Gunasena *et al.* (1978)	Maize	Sri Lanka	6 050	6 310
	Soyabean		940	950
Gunasena *et al.* (1978)	Maize	India	360	510
	Clusterbean		210	120

References

Ames, G. C. W. (1975). Peanuts: domestic, world production and trade. Research report 215, Dept of Agric. Econ., College Station, Athens, Georgia 30602, USA.

Brown, L. R. and Eckholm, E. P. (1975). "By Bread Alone", Pergamon, Oxford.

Byers, M. and Sturrock, J. W. (1965). The yields of leaf protein extracted by large-scale processing of various crops. J. Sci. Fd Agric. 16, 341–355.

Caldwell, B. E. (1973). "Soybeans: Improvement, Production and Uses", American Society of Agronomy, Madison, Wisconsin.

Evans, L. T. (1975). "Crop Physiology", Cambridge University Press, Cambridge.

F.A.O. (1954). Nutrition Studies No. 11. Food Composition Tables. F.A.O. Rome.

F.A.O. (1975). "Production Yearbook 28, 1974", F.A.O., Rome.

Gunasena, H. P. M., Campos, F. F. and Ahmed, S. (1978). Studies on intercropping and utilization of organic residues. 2nd INPUTS Review Meeting, East–West Center, Hawaii. May 1978.

Hardwick, R. C. and Milbourn, G. M. (1967). Yield analysis in the vining pea. Agric. Prog. 42, 24–31.

Heichel, G. H. (1973). Comparative efficiency of energy use in crop production. Bull. 739 Connecticut Agric. Expl Station, New Haven, U.S.A.

Heinrich, B. (1975). Energetics of pollination. In "Annual Review of Ecology and Systematics", Vol. 6, 139–170.

Hildreth, A. C., Magness, J. R. and Mitchell, J. W. (1941). Effects of climatic factors on growing plants. In "Yearbook of Agriculture 1941", U.S.D.A. Washington D.C.

I.C.R.I.S.A.T. (1975). International Workshop on Grain Legumes. I.C.R.I.S.A.T. Hyderabad.

Ishag, H. M. (1973). Physiology of seed yield in field beans. 1. Yield and yield components. J. Agric. Sci. Camb. 80, 181–189.

M.A.F.F. (1975). Tables of Feed Composition and Energy Allowances for Ruminants. H.M.S.O., London.

Martin, J. H., Leonard, W. H. and Stamp, D. L. (1976). "Principles of Field Crop Production", 3rd edn, Macmillan Publishing Co. New York and London.

McGregor, S. E. (1976). Insect Pollination of Cultivated Crop Plants. Agriculture Handbook No. 496. U.S.D.A. Washington, D.C., U.S.A.

Nix, J. (1976). "Farm Management Pocketbook", 7th edn, Wye College, Kent.

Oelshlegel, F. J., Schroeder, J. R. and Stahmann, M. A. (1969). Potential for protein concentrates from alfalfa and waste green plant material. J. Agr. Food Chem. 17, 791–795.

Pimental, D., Dritschilo, W., Krummel, J. and Kutzman, J. (1975). Energy and land constraints in food protein production. Science (N.Y.) 190, 754–61.

Protein Advisory Group of the United Nations System (1975). Nutritional Improvement of Food Legumes by Breeding. Wiley, New York.

Purseglove, J. W. (1968). "Tropical Crops: Dicotyledons I", Longmans, London.

Rachie, K. O. and Roberts, L. M. (1974). Grain legumes of the lowland tropics. Adv. Agron. 26, 1–132.

Rao, M. R. and Willey, R. W. (1978). Current status of intercropping research and some suggested experimental approaches. 2nd INPUTS Review Meeting at East–West Center, Hawaii.

Scott, R. K. (1974). Protein production from seed crops with particular reference to oilseed rape and field beans. In Proc. Reading University Agricultural Club Conference: The Production of More Homegrown Protein for Animal Feeding. (Eds. S. D. M. Jones and M. J. Walton-Evans).

Sheils, L. A. (1978). Personal communication.
Smartt, J. (1976). "Tropical Pulses", Longmans, London.
Snoad, S. (1974). A preliminary assessment of "leafless peas". *Euphytica* **23**, 257–265.
Snoad, B. and Davies, D. R. (1972). Breeding peas without leaves. *Span* **15**, 87–89.
Sutcliffe, J. F. and Pate, J. S. (1977). "The Physiology of the Garden Pea", Academic Press, London and New York.
Thompson, R. and Taylor, H. (1977). Yield components and cultivar sowing date and density in field beans. *Ann. Appl. Biol.* **86**, 313.
U.S.D.A. (1973). Agricultural Statistics 1973. Washington, D.C.
Willey, R. W. and Roberts, E. H. (1976). Mixed cropping. *In* Solar Energy in Agriculture. *Proc. Joint Conf.* University of Reading and U.K-I.S.E.S. Sept. 1976.

8

OIL CROPS

Plants produce a variety of oils and a considerable number of species are used, but only a few are of worldwide importance (see Table 8.1).

TABLE 8.1   Plant oils of world-wide importance

Edible		
Drying or semi-drying oil	Soyabean	*Glycine max*
Non-drying oil	Groundnut	*Arachis hypogeae*
Semi-drying oil	Sunflower	*Helianthus annuus*
Semi-drying oil	Rapeseed	*Brassica napus*
Semi-drying oil	Cottonseed	*Gossypium hirsutum*
Non-drying oil	Olive	*Olea europaea*
Non-drying oil	Sesame	*Sesamum indicum*
Edible – industrial		
Non-drying oil	Palm	*Elaeis guineensis*
Non-drying oil	Palm kernel	*Elaeis guineensis*
Non-drying oil	Coconut	*Cocos nucifera*
Industrial		
Drying oil	Linseed	*Linum usitatissmum*
Non-drying oil	Castor	*Ricinus communis*
Drying oil	Tung	*Aleurites fordii*

Source: Hartley (1977); Martin *et al*. (1976)

There are thirteen principal vegetable oils and oil seeds, two types of marine oil (whale and fish) and three categories of animal fat (butter, lard and tallow). Animal and marine oils and fats constitute some 25% of the total; the rest are of vegetable origin. The products vary from food to paint, soap and perfume, depending upon the kind of oil. Apart from the essential oils used for perfumes, most of them can be divided into the edible, non-drying oils (from palm, peanut, olive, almond, etc.), the semi-drying, inter-mediate oils (from cottonseed, maize, sesame, for example), and the drying oils (such as linseed). Most important oils come from seeds and most seeds contain some oil — sometimes over 50%: the residue after extraction is usually rich in protein. The major oil crops occupy some 70M ha, mainly in the tropics, but it is the fixed, non-volatile oils that are of greatest com-mercial importance.

Most oils are used in margarine, paint and soap but this could change. The use of such oils as sesame for cooking is of great antiquity but modern uses might be quite different. Castor oil, for example, once used widely in medicine, is now more likely to be used as a raw material for the production of sebacic acid from which synthetic resins and fibres can be manufactured.

Oils have very great advantages as products (incidentally, the distinction between oils and fats is a purely physical one, depending upon the melting point), quite apart from their common association with high protein resi-dues, and are often produced in large quantities per unit of land. The oil palm is considered to yield more oil per hectare than any other oil crop or, indeed, than the amount that can be produced per hectare from animal fat. The order of magnitude of these yields is given in Table 8.2.

The output of oil per unit of labour and support energy is illustrated in Table 8.3. The energetics of processing vary with the crop and whether by-products are dried artificially or by the sun.

The degree of processing required is also variable. For example, linseed oil is expressed from flaxseed by a variety of presses, sometimes in combina-tion with solvent extraction. When ground, cleaned linseed is pressed, about two-thirds of the oil may be squeezed out and most of the remainder can be extracted with solvent (Martin et al., 1976). The residual linseed cake contains between 1% and 6% of oil. Clearly, the efficiency of extraction could be varied, according to extraction costs and the need for oil in the resulting fractions.

The oil output can be fractionated for different purposes, often into fractions with different iodine numbers. The iodine number is a measure of the amount of oxygen that will be dissolved by the oil in drying to form a paint film. A high iodine number indicates a good drying quality for paint purposes; a low iodine number is desirable for hydrogenation to produce margarine.

Other methods of processing include steam distillation as used for remov-ing both peppermint and spearmint oil (used in confections, toothpaste, powders and perfumes).

Oil palms (*Elaeis guineensis*) give rise to three main products, palm oil,

TABLE 8.2   Yields of oil crops

Crop	Country	Fruit or seed yield (kg ha^{-1})	Oil content (%)	Oil content of crop (kg ha^{-1})	Extraction rate (%)	Oil yield (kg ha^{-1})
Soyabean	USA	1 620–2 700	13–20	210–540	95 (by solvent)	200–513
Groundnut	USA	1 350–1 688	38–50	513–844		
Sunflower	Argentina	1 000–2 000	20–40	200–800		
Rapeseed	Canada	1 680–2 800	40–46	672–1 288	34–38	228–489
Cottonseed	Egypt	1 340	15–25	181–284	13–18	24–51
Olive	USA	12 500	10–60	1 250–7 500		
Olive	Mediterranean	1 100–5 600		110–3 360		
Sesame	USA	930–2 240	50	465–1 120	35–50	163–560
Sesame	India	280–780		140–390		49–195
Oil palm (pericarp and kernel)	Sumatra (favourable conditions)					7 500
Oil palm (pulp)	Sumatra	600	56			2 200
Oil palm (kernel)	Sumatra	875–3 000			46–48	276–288
Coconut (copra)	—	630	64	564–1 935	62–99	352–1 916
Linseed	Canada	2 500	35–44	221–277	32–36	71–100
Castor	Brazil	3 360	35–55	875–1 375	90	788–1 238
Castor	USA	1 580–3 040	50–60	1 176–1 848		1 058–1 663
Tung	USA			790–1 824	31–37	245–675
		Milk yield				
Cows' milk	UK	6 665	3.5			233

Source: Godin and Spensley (1971)
Note: The tree crops — Olive, Oil palm, Coconut and Tung take a number of years to become productive and reach maturity, i.e. 6–20, 4–15, 6–20, 3–10, respectively.

TABLE 8.3   An illustration of the output of oil per unit of labour and support energy

(a) Labour[a]	Oil palms Malaysia 1 hour labour yields 8 kg oil
(b) Support energy[b]	Safflower California, USA 87 MJ support energy yields 8 kg oil

[a]Source: Hartley (1977); Godin and Spensley (1971)
[b]Source: Calculated from Cervinka et al. (1974)

palm kernel oil and palm kernel cake. The last has a relatively low protein content (18–19%) and is thus more suitable for ruminants. Palm kernels contain about 47–52% oil, 6–8% moisture, 7.5–9.0% protein, 23–24% extractable non-protein, 5% cellulose and 2% ash (Hartley, 1977). Extraction methods vary from the non-mechanical traditional methods, for soft oil production or hard oil production (with extraction efficiencies of 40–45% and 20%, respectively), to power-operated mills of various kinds, operating at an efficiency of extraction (i.e. extracted oil to total oil in the fruit) of 87–88%. The advantages of palm oil are mainly versatility in use and high yield of oil per hectare: the main disadvantage is the relatively high extraction cost. Palm and palm kernel oils together represent about 17% of all vegetable oil production and c. 13% of all oils and fats.

The support energy used in growing a crop is greatly influenced by the fertilizer requirement, particularly for nitrogen. Protein output per unit of nitrogen applied varies primarily according to whether crops are leguminous or not (Table 8.4).

Oilseeds, especially peanuts and soyabeans, are important in world trade but a high proportion of the products are destined for animal feed. The efficiency with which protein is produced in this way is, of course, very much lower than for direct consumption. Whereas forage crops cannot usually be utilised directly for human consumption, the products of oilseeds (such as soyabeans) can and are so used.

As Pyke (1970) has pointed out, at one time much of the oil and fat produced was used to make candles or as fuel for lamps. The advent of electricity released these oils for direct consumption. In some cases, there are problems of palatability, and even the need to eliminate toxic compounds, and it may be of increasing importance that the processing required involves the use of energy.

TABLE 8.4   Protein output by oil crops per unit of nitrogen fertilizer applied

Crop	N use (kg ha^{-1})	Grain yield (kg ha^{-1})	Protein content (%)	Protein yield (kg ha^{-1})	Protein yield (kg) per kg fert. N used
Soyabeans (USA)[a]	22	1836	25–52	459–955	21–43
Peanuts (USA)[a]	22	2360	25–34	566–802	26–36
Oil seed rape (UK)	230	2000	17.2	344	1.5
Castor oil seed (USA)	135	3360	21.0	706	5.2
Safflower (USA)	84–112	2800–4480	15.4–22.5	431–1 008	5.1–9
Sesame (USA)	40–90	930–2240	19–25	177–560	4.4–6.2

[a]see Table 7.9

Sources: Weiss (1971); Nix (1976)

Even so, at a time when substitutes for fossil fuels are being sought, oil-producing crops are bound to be of great interest. Not only do they include some very productive crops, capable of making good use of solar radiation, they also contain a readily usable energy source as expressible oil. Fuel crops are being explored that produce large quantities of biomass but the processing of large quantities of material, often containing a lot of water, poses some problems and may not always lead readily to a liquid fuel. Yet there are many needs for liquid fuels and a direct crop source has advantages.

However, no crop produces only oil and, indeed, the main oil crops are very good examples of long-established multiple use. The fractionation of products and by-products derived from the oil palm fruit is illustrated in Fig. 8.1.

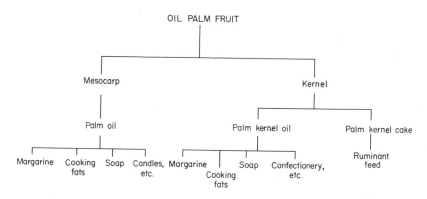

FIG. 8.1   The fractionation of products and by-products derived from the oil palm fruit (after Hartley, 1977)

Clearly, the overall efficiency of the total production process is influenced by the efficiency of the constituent processes but for the most part these do not greatly interact.

References

Cervinka, V. Chanellor, W. J., Coffett, R. J., Curley, R. G. & Dobie, J. B. (1974). "Energy Requirements for Agriculture in California", University of California and California Dept. of Food & Agriculture.
Godin, V. J. and Spensely, P. C. (1971). Oils and Oilseeds. Tropical Products Institute Crop and Products Digests No. 1. T.P.I., London.
Hartley, C. W. S. (1977). "The Oil Palm", 2nd edn, Longmans, London.

Martin, J. H., Leonard, W. H. and Stamp, D. L. (1976). "Principles of Field Crop Production", 3rd edn, Macmillan, New York and London.
Nix, J. (1976) "Farm Management Pocketbook", 7th edn, Wye College, Ashford.
Pyke, M. (1970). "Man and Food", World University Library, London.
Weiss, E. A. (1971). "Castor, Sesame and Safflower", Leonard Hill, London.

9

ROOT AND TUBER PRODUCTION

Roots and tubers are grouped together here simply because they are all underground storage organs. They may usefully be compared with seed-producing plants (Chapters 6 and 7) and with foliage producers (Chapter 10), in terms of their agricultural productivity and of the kinds of land on which they can be grown.

Many species of root crop are grown (Table 9.1) but few of them contribute significant quantities of food to the world total. Locally they may be important producers of animal feed, partly because they can be simply stored for this purpose, and some root crops, such as cassava, are staple foods in some parts of the world (Coursey and Haynes, 1970). Many are grown as vegetables or salad crops and some need processing before they can be consumed, in order to remove or alter toxic substances they contain (Oke, 1968). This is true of the most important of the world root crops, cassava or manioc (*Manihot esculenta*). The need to process and prepare is not confined to root crops, of course, and it can have a considerable influence on the output of edible product and thus on many aspects of efficiency. This is illustrated in Table 9.2 for cereals, roots and tubers, in relation to preparatory processes that may be regarded as normally essential. The rather special case of sugar beet is dealt with in Chapter 12.

All this takes no account of cooking, and considerable further losses may occur during this phase. However, the degree and extent to which cooking is carried out is extremely variable and many cooking processes (such as boiling potatoes) are aimed at rendering the product better in some nutritional sense (more digestible, for example). Some of these aspects are illustrated in Table 9.3. Improvements due to processing will only be reflected in efficiency ratios if these are based on nutritionally qualified outputs, such as digestible or metabolizable energy (or nitrogen), rather then gross nutrient content. This can make a big difference (see Table 9.4) but, since it often reflects processing, is hardly an expression of the plant products

105

themselves. In the main, then, we will adhere to gross nutrient output as a basis of productivity calculations.

TABLE 9.1   Important root crops

	World Production 1975 (1000 metric tonnes fresh wt)	Percentage of total world production of roots and tubers[a]
Starchy roots		
Cassava (*Manihot esculenta*)	105 209	12.9
Yams (*Dioscorea* spp.)	20 198	2.5
Other roots		
Sugar beet (*Beta vulgaris*)	249 851	30.6
Vegetables		
Carrot (*Daucus carota*)	5 356	0.7
Swedes or Rutabaga	NA[b]	
(*Brassica napus* var *napobrassica*)		
Turnip (*Brassica campestris* var. *rapa*)	NA	
Radish (*Raphanus sativus*)	NA	

[a]Total world production of roots and tubers is taken here as the totals in this table and Table 9.9 plus the total given by F.A.O. for minor root and tuber crops giving a total of 817 446 000 metric tonnes. Starchy roots and tubers provide 7% of Man's energy sources, sugar from cane and beet a further 7%. Starchy roots and tubers provide 5% of Man's protein sources.
[b]NA — not available in F.A.O. production tables

Sources: F.A.O. (1976); Brown and Eckholm (1975); Evans (1975a)

The gross output of energy per hectare can be very high from root crops (Table 9.5), although they tend to have a high water content (Table 9.6) and are thus costly to handle and transport. Fortunately, they can often be stored for long periods, many of them left in the ground until required. Their protein yields per unit of land, however, are not generally high (Table 9.7) and their protein contents are normally low (Table 9.8).

Root crops most commonly have a single enlarged tap-root (turnip, swede, beet, radish, carrot, parsnip) but sometimes the root structure is modified to form root tubers (as in cassava). The main tuber crops are listed in Table 9.9, from which it can be seen that, in terms of their contribution to world food supplies, they are dominated by the potato and the sweet potato. Where they can be grown, such crops can produce very high yields per

TABLE 9.2 The effect of processing and preparation on roots, tubers and cereals

Crop	Method of processing or preparation	Percentage of weight lost in processing	Source
Potatoes	Peeling	13.5	Chappell (1954)
	Scraping	2.8	Chappell (1954)
Carrots	Peeling and topping	14.5	Chappell (1954)
Yams	Peeling	10.0–15.0	Kay (1973)
		23.0	Francis et al. (1975)
Cassava	Peeling	10.0–20.0	Kay (1973)
		23.0	Dalton and Akwetey (1971)
Rice	Milling	10.0	Inglett (1975)
Wheat	Milling — White flour	27.7	Roy (1976)
	— Brown flour	10.0–20.0	Roy (1976)
	— Wholemeal	nil	Roy (1976)
Maize	De-husking	5.0	Dimler (1960)
	Milling	20.0	Dimler (1960)
Oats	Milling	25.0	Inglett (1975)

TABLE 9.3 The effect of cooking on nutritional value of roots and tubers (per 100 g edible material except where stated)

Crop	Method of cooking	Protein content (g)[a]		Gross Energy content (kJ)[a]	
		Raw	Cooked	Raw	Cooked
Potatoes	Boiling	2.1	1.4	318	331(1)
	Frying[b]	2.1	3.8	318	989(1)
Carrots	Boiling	0.7	0.6	105	84(5)
Yams	Boiling	1.79(2)	1.79(2)	435(3)	418–502(4)
Sweet potato[c]	Boiling	1.4	1.1	423	364(5)

[a]Sources: (1) MAFF (1970); (2) Francis et al. (1975); (3) Platt (1962); (4) Coursey (1967); (5) Mottram and Radloff (1937).
[b]Fat is added
[c]Values for raw material are as purchased

TABLE 9.4   Coefficients of apparent digestibility and nitrogen retention for potato-based pig diets (after Whittemore *et al.*, 1973)

	Digestibility of gross energy	Digestibility of nitrogen	Nitrogen retention (g per 100 g digestible N)
Cooked potato flour fed alone	0.961	0.89	37.1
Raw potato fed alone	0.915	0.69	−40.1
Cooked potato fed with a maize balancer meal	0.922	0.90	57.9
Raw potato fed with a maize balancer meal	0.877	0.80	26.4

[a]Represents high losses of nitrogen in urine

hectare, especially of gross energy. Total protein production may be considerable but contents are low (Table 9.10) and, because much of the protein is in or near the skin, it may be lost in preparation or processing (Table 9.11).

Whereas root crops are mostly grown from seed and the weight of seed required is very small (Table 9.12), tuber crops are generally grown from tubers (although these can be subdivided) and quite a high proportion of the yield may have to be devoted to sowing the next crop (Table 9.13). In neither case, however, is it usually as simple as that. Root crops, as such, do not produce seed and special crops have to be grown on for a second year to do so. For different reasons, related to disease resistance, tuber crops, such as potatoes for "seed", may also be grown quite separately. It is therefore more accurate to express the areas devoted to "seed" production as a proportion of those sown (Table 9.14), in order to obtain a measure of the resources that have to be devoted to ensuring the next crop.

Both roots and tubers contain considerable proportions of water and crop yields are very sensitive to drought. Response to irrigation depends, of course, on the extent of the water shortage, but the water needs of these crops are substantial (Table 9.15), though not as markedly different from say, cereals as might be expected.

In most circumstances, they are also regarded as labour demanding (Table 9.16), although cassava is highly productive per unit of labour

TABLE 9.5 Output of gross energy from root crops

Crop	Yield (kg ha^{-1})	DM (%)	Yield (kg DM ha^{-1})	Energy content (MJ kg^{-1})	Gross energy output (GJ ha^{-1})
Cassava[a]	3 000–50 000	38	1 140–19 000	4.56 (fresh material)	14–228
Yam	7 500–30 000	28	2 100– 8 400	3.77 (fresh material)	28–113
Sugar beet (UK)	28 000–44 000	23	6 440–10 120	17.6 (DM)	113–178
Carrot (UK)	25 000–35 000	13	3 250– 4 550	17.4 (DM)	57–79

[a]Growth period is 9–24 months, when used as vegetables they are normally harvested within 12 months

See Table 6.4, p. 78, for outputs of energy by grains

Sources: cassava and yam yields — Kay (1973); cassava and yam energy contents — Chatfield (1954); sugar beet and carrot yields — Nix (1976); sugar beet and carrot energy contents — M.A.F.F. (1975)

TABLE 9.6   Water content of root crops

Crop	Water content (%)
Cassava (peeled)	62–65
Yams	60–70
Sugar beet	77
Carrots	87
Swedes or rutabaga	88
Turnips	91
Radish	94

See Table 6.2, p. 76, for water content of grains

Sources: Kay (1973); M.A.F.F. (1975)

(Coursey and Haynes, 1970). Fertilizer needs are relatively high: the quantities normally applied per kg of product are shown in Table 9.17.

As with other crops, if the fertilizer input is inadequate, the efficiency with which all other resources are used tends to be reduced: even so, it is always possible to apply more than is necessary or to apply it at times when it is inefficiently used or simply wasted (by leaching, for example). Usually, the most complicated effect is on the efficiency of support energy use. That part of the support energy that is not supplied in the form of fertilizer is, like the other resources, used more efficiently as the level of fertilizer application increases, up to an optimum quantity. Since the fertilizer itself may account for such a large part of the total support energy input, the efficiency with which the latter is used tends to improve so long as the use of additional fertilizer continues to improve yield (Table 9.18). Typical patterns of support energy use in root and tuber crop production are given in Table 9.19.

Examples of efficiency of support energy use by these crops are given in Table 9.20, for both gross energy and protein production.

Factors influencing efficiency

The fundamental processes that determine yield are the growth of the whole plant and the partition of nutrients to the storage organ. There is no ripening phase and time tends to be less of a constraint, either in terms of when growth occurs or when the product is harvested. There are, however, quite critical phases in establishing the crop, especially in temperate climates

TABLE 9.7   Output of protein from root crops

Crop	Yield (fresh material) (kg ha⁻¹)	DM (%)	Yield (kg DM ha⁻¹)	Protein content (%)	Protein output (kg ha⁻¹)
Cassava	3 000–50 000	38	1 140–19 000	0.9 (fresh material)	27–450
Yam	7 500–30 000	28	2 100– 8 400	2.1 (fresh material)	158–630
Sugar beet	28 000–44 000	23	6 440–10 120	4.8 (DM)	309–486
Carrot	25 000–35 000	13	3 250– 4 550	9.2 (DM)	299–419

See Table 7.3, p. 88, for output of protein by pulse crops

Sources: yields as for Table 9.5; protein contents for cassava and yam — Chatfield (1954); protein contents for sugar beet and carrot — M.A.F.F. (1975)

TABLE 9.8   Protein content of root crops

Crop	Protein content of fresh material (%)	Protein content of DM (%)
Cassava	0.9	2.4
Yams	2.1	7.6
Beet	1.3	10.5
Carrot	1.0	7.7
Swede	0.9	7.5
Turnip	0.9	10.0
Radish	0.6	10.0

Source: Chatfield (1954)

TABLE 9.9   Important tuber crops

	World production 1975 (1 000 metric tonnes fresh wt)	Percentage of total world production of roots and tubers[a]
Potato (*Solanum tuberosum*)	291 321	35.6
Sweet potato (*Ipomoea batatas*)	136 570	16.7
Jerusalem artichoke (*Helianthus tuberosus*)	NA	

[a]See Table 9.1 for explanation and sources

TABLE 9.10   Protein content of tuber crops

Crop	Protein content of fresh material (%)	Protein content of DM (%)	Source
Potato	1.7	7.7	(1)
	0.7–4.6	1.9–35.1	(2)
Sweet potato	1.1	3.6	(1)
	0.95–2.4	1.9–12.6	(2)
Jerusalem artichoke	1.5	7.5	(1)
	0.9–3.25	3.2–20.6	(2)

Sources: (1) Chatfield (1954); (2) Kay (1973)

TABLE 9.11 Protein content of potatoes (after Bacharach and Rendle, 1946)

	Percentage crude protein in dry matter
Whole potatoes	8.82
Peelings (by hand)	9.94
Peelings (by machine)	9.15

TABLE 9.12 Weight of seed sown for root crops

Crop	Wt of seed planted (kg ha^{-1})
Sugar beet (UK and USA)	1–9
Beet	13.5–51.3
Carrot	1.4–4.7
Swede UK	1.2–2.1
Turnip	2.0–3.3
Parsnip	1.7–3.0

Sources: Martin et al. (1976); M.A.F.F. (1977); Elsoms' seed catalogue.

TABLE 9.13 Proportion of tuber crops required for propagation (after Kay, 1973)

Crop	Crop yield (kg ha^{-1})	Wt req'd for propagation (kg ha^{-1})	Proportion (%)
Potato	5 000– 42 500	1 255–1 880	25–4
Sweet potato[a]	1 250– 50 000	382–470	30–1
Jerusalem artichoke	12 000–160 000	320–553	3–0.3

[a]There are two methods of propagation for sweet potatoes in general use: vine cuttings, in which case none of the tubers are used; or transplants, where tubers are planted, usually in a nursery bed, and then transplanted when the shoots have grown.

TABLE 9.14   Areas required for "seed" production for root and tuber crops

Crop	Yield of "seed" (kg ha^{-1})	"Seed"req'd for planting (kg ha^{-1})	Area req'd (ha) for "seed"production to plant 1 ha
Beet	2 130 879	13.5–51.3	0.012–0.04
Carrot	785 251–628	1.4– 4.7	0.004–0.012
Parsnip	897 1 256–1 883	1.7–3.0	0.004–0.06
Radish	1 009	24.5–61.1	0.02 –0.06
Rutabaga (swede)	637 1 785–2 009	1.2– 2.1	0.004–0.006
Turnip	637 1 256–1 758	2.0– 3.0	0.006–0.008
Potato[a]	20 000–36 000	2750	0.08 –0.14

NB Apart from the radish and potato the root crops listed are biennial and require two seasons to produce "seed". Therefore, the area allowed is for both stages of production.

[a]This calculation assumes that the whole crop could be used for "seed" and is grown as a normal maincrop. In fact usually only part of the crop is suitable for "seed" and for health reasons may be harvested early with a resulting lower yield

Sources: Nix (1976); Shoemaker (1953); Bowring (1978)

where there is a limited period between the last spring frosts and the peak of incident solar radiation. It is necessary to establish a full plant population in time to take advantage of the available radiation. There are thus two main phases: one, to get plants established and to achieve an adequate leaf area index; two, to ensure a good supply of nutrients during the development of the storage organs.

TABLE 9.15 Water requirements for root, tuber and grain crops

Crop	Water req'd[a]	Source
Potato	499	(1)
Potato	650	(1)
Potato	636	(2)
Turnip	614	(1)
Sugar beet	377	(1)
Sugar beet	397	(2)
Millet	308	(1)
Sorghum	309	(1)
Maize	325	(1)
Wheat	503	(1)

[a]g of water used per g of plant DM produced

Sources (1) Maximov (1929); (2) Hilldreth et al. (1941)

TABLE 9.16 Labour requirements

Crop		Yield (kg ha^{-1})	Labour (man h ha^{-1})	Source
Potato	(USA)	26 208	60	(1)
Potato	(USA)	29 003	114	(2)
Potato	(UK)	25 000–30 000	59–155	(3)
Potato	(Fiji)	7 168[a]	517	(4,5)
Sugar beet	(USA)	50 221	79	(2)
Sugar beet	(UK)	36 000	36–73	(3)
Carrot	(UK)	30 000	105	(3)
Cassava	(Fiji)	8 489[a]	467	(4,5)
Sweet potato	(Fiji)	7 580[a]	358	(4,5)
Maize	(USA)	5 080	22	(1)
Rice	(USA)	5 796	30	(1)
Wheat	(USA)	2 284	7	(1)

[a]Edible, not gross, yield

Sources: (1) Pimental et al. (1975); (2) U.S.D.A. (1973); (3) Nix (1976); (4) Chandra et al. (1974); (5) Chandra et al. (1976)

TABLE 9.17   The output of root and tuber crops per unit of applied fertilizer nitrogen

Crop	N input (kg ha^{-1})	Yield (fresh wt) (kg ha^{-1})	Output/kg N (kg)
Cassava	54–80	3 000–50 000	38–926
Sweet potato	34–45	2 500–50 000	56–1470
Carrot (UK)	60–80	25 000–35 000	312–583
Potato (UK)	120–200	20 000–36 000	100–300
Swede or turnip (UK)	60–100	32 500	325–541

Sources: cassava and sweet potato — Kay (1973); carrot, potato and swede or turnip — nitrogen usage, Eddows (1976), yields, Nix (1976), M.A.F.F. (1977)

TABLE 9.18   Effect of nitrogen level on support energy usage in potatoes

System	Nitrogen application (kg ha^{-1})	Yield of tubers (kg ha^{-1})	Support energy used (MJ per kg tubers)	Energy ratio (Gross energy in tubers/Support energy input)
Potato Marketing Board statistics 1975[a] UK Maincrop	138	22 300	2.30	1.38
Good UK production[b]	170	27 500	1.89	1.69
Blueprint potatoes[c]	250	87 000	1.33	2.40

[a]Potato Marketing Board (1976)
[b]White (1975)
[c]Evans (1975b)

TABLE 9.19   Typical patterns of support energy inputs (MJ tonne⁻¹) in root and
tuber crop production

	Cassava(1) in Australia		Carrots(2) in UK
	Tops	Tubers	Roots
Fuel for cultivation	142	142	104
Fuel for harvesting	140	414	} 92
Transportation	133	147	
Machinery manufacture and maintenance	81	230	41
Herbicides/fertilizers	350	350	184
Material for replanting	28	28	—
Total	874	1 311	421

Sources: (1) Saddler (1975); (2) Shiels (1978)

TABLE 9.20   Examples of the efficiency of support energy use by roots and
tubers

	Gross energy yield (MJ) / Support energy input (MJ)	Crude protein yield (kg) / Support energy inputs (MJ)
1. *Subsistence crops in Fiji*(1)		
Cassava	71	0.14
Sweet potato	82	0.22
Yams	77	0.43
2. *Commercial crops in UK*		
Potatoes(3)	1.7	0.011
Carrots(2)	2.3	0.017
Beetroot(2)	1.9	0.018
Parsnips(2)	2.7	0.022

Sources: (1) after Chandra *et al.* (1974) and Chandra *et al.* (1976); (2) Shiels (1978); (3)
White (1975)

References

Bacharach, A. L. and Rendle, T. (1946). "The Nation's Food — A Survey of Scientific Data", Society of Chemical Industry, London.

Bowring, J. D. C. (1978). National Institute of Agricultural Botany, personal communication.

Brown, L. R. and Eckholm, E. P. (1975). "By Bread Alone", Pergamon, Oxford.

Chandra, S., De Boer, A. J. and Evenson, J. P. (1974). Economics and energetics; Sigatoka Valley, Fiji. *World Crops* **26(1)**, 34–7.

Chandra, S., Evenson, J. P. and De Boer, A. J. (1976). Incorporating energetic measures in an analysis of crop production practices in Sigatoka Valley, Fiji. *Agric. Systems* **1(4)**, 301–311.

Chappell, G. M. (1954). Food waste and loss of weight in cooking. *Br. J. Nutrition* **8(4)**, 325–340.

Chatfield, C. (1954). Nutrition Study No. 11. Food Composition Tables. F.A.O. Rome.

Coursey, D. G. (1967). "Yams", Longmans, London.

Coursey, D. G. and Haynes, P. H. (1970). Root crops and their potential as food in the tropics. *World Crops* **22**, 261–265.

Dalton, G. E. and Akwetey, F. (1971). Cassava production and processing in South East Ghana. Department of Agric. Economics Farm Management, University of Ghana.

Dimler, R. J. (1960). Effects of commercial processing of cereals on nutrient content. *In* "Nutritional Evaluation of Food Processing", (Eds. R. S. Harris and H. Von Loesecke), Wiley, New York.

Eddowes, M. (1976). "Crop Production in Europe", Oxford University Press, Oxford.

Evans, L. T. (1975a). "Crop Physiology", Cambridge University Press, Cambridge.

Evans, S. A. (1975b). Maximum potato yield in the United Kingdom. *Outl. Agric.* **8(4)**, 184–7.

F.A.O. (1976). Production Yearbook 1975. Vol. 29. F.A.O., Rome.

Francis, B. J., Halliday, D. and Robinson, J. M. (1975). Yams as a source of edible protein. *Tropical Sci.* **17(2)**, 103–110.

Hilldreth, A. C., Magness, J. R. and Mitchell, J. W. (1941). Effects of climatic factors on growing plants. Yearbook of Agriculture 1941. U.S.D.A., Washington D.C.

Inglett, G. E. (1975). Effects of refining operations on cereals. *In* "Nutritional Evaluation of Food Processing", 2nd edn, (Eds R. S. Harris and E. Karmas), AVI Publishing Co., Westport, Connecticut.

Kay, D. E. (1973). "Root Crops", Tropical Products Institute Crop and Product Digest No. 2. Tropical Products Institute, London.

M.A.F.F. (1970). "Manual of Nutrition", H.M.S.O., London.

M.A.F.F. (1975). Tables of Feed Composition and Energy Allowances for Ruminants. M.A.F.F., Pinner.

M.A.F.F. (1977). Output and utilization of farm produce in the U.K. 1969/70 to 1975/76.

Martin, J. H., Leonard, W. H. and Stamp, D. L. (1976). "Principles of Field Crop Production", 3rd ed, Macmillan, New York and London.

Maximov, N. A. (1929). "The Plant in Relation to Water", George Allan and Unwin, London.

Mottram, V. H. and Radloff, E. M. (1937). "Food Tables", Arnold, London.
Nix, J. (1976). "Farm Management Pocketbook", 7th edn, Wye College, Kent.
Oke, O. L. (1968). Cassava as food in Nigeria. *World Rev. Nutr. Dietetics* **9**, 227–250.
Pimental, D., Dritschilo, W., Krummel, J. and Kutzman, J. (1975). Energy and land constraints in food protein production. *Science (N.Y.)* **190**, 754–61.
Platt, B. S. (1962). Tables of representative values of foods commonly used in tropical countries. M.R.C. Special Report Series No. 302. H.M.S.O., London.
Potato Marketing Board (1976). Maincrop potato production in Great Britain 1975–76., Potato Marketing Board, Oxford.
Roy, R. (1976). "Wastage in the UK Food System", Earth Resources Research, London.
Saddler, H. D. W. (1975). "Organic Wastes and Energy Crops as Potential Sources of Fuel in Australia", The Energy Research Centre, University of Sydney, Sydney, Australia.
Shiels, L. A. (1978). Personal communication.
Shoemaker, J. S. (1953). "Vegetable Growing", 2nd edn, Wiley, New York.
U.S.D.A. (1973). Agricultural Statistics, 1973. Washington, D.C.
White, D. J. (1975). Energy in agricultural systems. Annual Conference, Inst. Agric. Engineers, London.
Whittemore, C. T., Taylor, A. G. and Elsley, F. W. H. (1973). The influence of processing upon the nutritive value of the potato: Digestibility studies with pigs. *J. Sci. Fd Agric.* **24(5)**, 539–545.

10

LEAVES, STEMS AND FLOWERS
(INCLUDING BULBS AND CORMS)

"Leafy crops" covers a wide variety of crop species, some of them rather unrelated to each other in morphology and function. For example, the tea plant (*Thea sinensis*) is undeniably a leafy crop, the leaves of which form a commercial product of some importance. It is so different agriculturally, however, from the grasses and brassicas, for example, that it is most conveniently dealt with separately (in Chapter 13).

Even so, we have still to include browse trees and shrubs, pastures, sown forage crops (such as rape and kale), whole-crop cereals and pulses grown for conservation, vegetables for both direct human consumption and for animal feed, and salad crops.

Although crops grown for animal feed and those grown for direct human consumption may seem very different and possibly not worth comparing, it is worth remembering that most plant products are processed, one way or another, before human consumption and that the differences between processing by the animal and processes involving, for example, crushing, fermentation and extraction, are not fundamentally very great. Furthermore, those that are normally fed to animals could in the future be processed in some other way (for example, by leaf protein extraction or green crop fractionation, as it is now generally called). One object of comparisons of various aspects of efficiency might be to determine whether they *should* be processed in a different way.

The present importance of "leafy crops" is difficult to assess. The contribution of leafy vegetables and salads to the world's food supply is not a major one (see Table 10.1), although it is hard to estimate with any accuracy the quantities grown and consumed by the grower and his family without entering into any kind of trade, even for a country like the UK. It is also true that the significance of a particular food may not be well expressed by the quantity produced or consumed. In fact, some leafy vegetables are greatly

120

TABLE 10.1 The contribution of various food groups to average world daily per caput intake (after F.A.O., 1971)

	Energy		Protein	
	(KJ)	(%)	(g)	(%)
Cereals	5 209	52.4	31.1	47.4
Animal products	1 347	13.6	20.7	31.5
Roots and tubers	770	7.8	2.8	4.3
Sugar and sugar products	879	8.8	0.1	0.2
Pulses, nuts, oilseeds	506	5.1	7.9	12.0
Vegetables[a]	151	1.5	2.2	3.4
Fruits	197	2.0	0.6	0.9

[a]This includes other vegetables besides the leafy ones.

valued for their protein content and many more for the minerals and vitamins that they contain (Rajalakshmi and Ramakrishman, 1969; Schmidt, 1971; Shepherd, 1975).

Grassland, on the other hand, is usually regarded as one of the world's most important crops and it certainly occupies a substantial proportion of the agricultural areas of the world (Table 10.2). Much of this land, however, achieves only a very low level of primary productivity and the further conversion to animal products is often extremely inefficient (see Table 10.3). The result is that many grassland areas are important socially, to people who may totally depend upon them, but their output of human food is very small compared with comparable arable areas. Of course, it also has to be remembered that much of the world's grassland is wholly or partly unsuitable for arable use, due to harsh climate, difficult terrain or poor soil,

TABLE 10.2 Proportion of the world's agricultural land occupied by grassland (after F.A.O., 1974)

	Agricultural land	
	(10^6 ha)	(%)
Arable land and land under permanent crops	1 475	33
Permanent meadows and grassland	3 005	67

TABLE 10.3   Efficiency of conversion of grass to animal products

$$\text{Efficiency} = \frac{\text{Energy output in product}}{\text{Energy intake as grass or grass products}}$$	
Lamb carcase from Kerry Hill ewe with single Suffolk cross lamb	0.065
Beef carcase from weaned Friesian cross calves	0.052
Milk from Friesian cow	0.256

Sources: Tayler (1970); Walsingham *et al*. (1975); Le Du (1976)

and that the animal products derived from them are not limited to food (see Chapters 25, 26 and 27).

There is one major difference in the assessment of efficiency where leafy crops are consumed in the fresh state by animals. This relates to the importance of the pattern, or seasonality, of production. If forage is dried or conserved as silage, the problem is no different from that of feeding grain or stored roots, for example. This may even be so for the grazed situation, where grass may remain virtually as standing hay during a dry season.

Under most grazing conditions, however, efficiency of animal production (per unit of land or per unit of feed) will be greatly influenced by the pattern of feed supply. This will be further dealt with in Chapter 18, but the importance of pattern of growth and production of leafy crops for grazing should be recognized at this point in the discussion. Typical production curves for grasses and legumes are shown in Figs 10.1 and 10.2

Pasture Grasses and Legumes

The productive efficiency of forage grasses and legumes tends to reflect their ability to grow in a particular environment. Examples of production of energy and protein by temperate grasses and legumes, per hectare, per megajoule of solar radiation and per megajoule of support energy, are given in Tables 10.4 and 10.5. Comparable figures for tropical situations (Table 10.6) reflect, amongst other things, the fact that the most important group of pasture legumes (the clovers — *Trifolium* spp.) only flourish in the cooler areas. *Trifolium subterraneum* (subterranean clover), for example, for all its enormous contribution to agriculture, is not a truly tropical species and, near the equator, it is confined to altitudes over 2400 m.

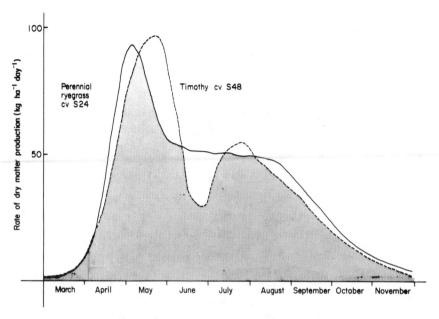

FIG. 10.1   Seasonal profiles of dry matter production for two contrasting grasses (from Green and Corrall, 1974)

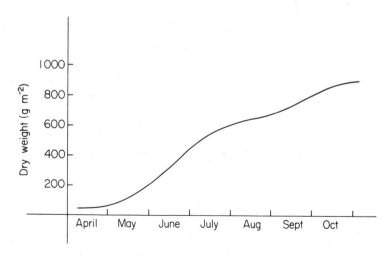

FIG. 10.2   The calculated course of dry weight increase in undefoliated swards of subterranean clover, 1950 (in Adelaide) (after Black, 1964)

TABLE 10.4 The production efficiency of temperate grasses (calculated from McBratney and Laidlaw, 1974)

	Yield of gross energy in product (MJ)			Yield of protein in product (kg)		
	per ha	per MJ solar radiation	per MJ[a] support energy used	per ha	per MJ solar radiation	per MJ[a] support energy used
Perennial ryegrass	263 625	0.0066	3.15	2 750	0.0001	0.030
Cocksfoot	243 830	0.006	2.95	2 727	0.0001	0.028
Tall fescue	262 330	0.007	3.26	2 667	0.0001	0.031

[a]When silage is made

TABLE 10.5 The production efficiency of temperate legumes (calculated using unpublished data from The Grassland Research Institute, Hurley, Berkshire)

	Yield of gross energy in product (MJ)			Yield of protein in product (kg)		
	per ha	per MJ solar radiation	per MJ support energy used	per ha	per MJ solar radiation	per MJ support energy used
Lucerne	195 800	0.005	6.40	2 024	0.00005	0.056
Red clover	151 799	0.0038	4.65	1 592	0.00004	0.032

TABLE 10.6   Production efficiency of tropical grasses and legumes

	Yield of protein (kg) in product	
	per ha	per GJ solar radiation
Napier grass[a]	195	0.003
Guinea grass[a]	141	0.002
Para grass[a]	122	0.002
Stylosanthes guianensis (Stylo, Tropical Lucerne)[b]	1 681	0.03
Desmodium intortum (Green leaf desmodium)[c]	2 634	0.04
Glycine wightii[d]	697	0.01

[a]Vincente-Chandler et al. (1959) Grasses grown in different soils over different periods with nitrogen levels somewhat higher than used in practice)   [b]Gilchrist (1967)   [c]Whitney et al. (1967)   [d]Colman et al. (1966)

The legumes tend to have a higher protein content than the grasses (Table 10.7) and, since they usually require no added nitrogen, they tend to be very efficient producers of protein per unit area of land and inputs. Some of the deep-rooting legumes (e.g. lucerne, *Medicago sativa*) are also capable of performing well under relatively dry conditions. The quantities of water

TABLE 10.7   The protein[a] content of grasses and legumes (after Spedding and Diekmahns, 1972)

	Percentage protein in DM
Perennial ryegrass	6.9–29.4
Cocksfoot	8.8–28.8
Timothy	9.4–28.8
Sainfoin	12.5–28.5
Lucerne	12.9–26.9
Red clover	15.6–26.1
White clover	21.1–32.6

[a]N content × 6.25

used in the production of pasture grasses and legumes are indicated in Table 10.8.

TABLE 10.8   Water requirements of pasture grasses and legumes

Crop	Water[a] requirement	Sources[b]
Buffalo grass (*Bulbilis dactyloidis*)	296	(1)[b]
Grama grass (*Bouteloua gracilis*)	338	(1)
Wheat-grass (*Agropyron desertorum*)	678	(1)
Wheat-grass, western (*Agropyron smithii*)	1 035	(1)
Awnless brome-grass (*Bromus inermis*)	977	(1)
Clover, sweet (*Melilotus alba*)	731	(1)
Clover, sweet	770	(2)
Clover, red (*Trifolium pratense*)	759	(1)
Clover, crimson (*Trifolium incarnatum*)	606	(1)
Clover	797	(2)
Alfalfa (*Medicago sativa*)	626–920	(1)
Alfalfa	831	(2)
Tropical grasses	320	(3)
Temperate grasses	670	(3)

[a]Grams of water used per g of plant DM produced
[b]Sources: (1) Maximov (1929); (2) Hilldreth *et al.* (1941); (3) Downes (1969)

Legumes are generally higher in mineral content than grasses but, when considering the chemical constituents of leafy crops, the enormous variation that can be found should be taken into account. Some examples of this variation are given in Table 10.9 for the mineral composition of grasses and legumes.

Labour usage on pasture tends to be relatively low in relation to production, although this is sometimes difficult to distinguish from the labour involvement in utilization. In perennial crops the initial cultivations are spread over several years and a great deal of output and the biggest labour input is for fertilizer application and harvesting for conservation.

Fertilizer manufacture and application, and harvesting for conservation usually represent the largest inputs of support energy (Table 10.10) unless the crop is artificially dried.

Efficiency of support energy use therefore tends to reflect the fertilizer input and the method of conservation. Artificial drying is very expensive in energy terms but reduces losses, quantitatively and qualitatively, and therefore increases output per unit of land and applied fertilizer. Fertilizers are also costly but, if sensibly applied, may increase yield in a linear fashion. The

TABLE 10.9 Variation in the mineral composition (% of DM) of grasses and legumes (after Spedding and Diekmahns, 1972)

	N	P	K	Mg	Na	Mn	Cu
Perennial ryegrass	1.1 –3.79	0.18–0.5	1.45–3.08	0.07–0.37	0.04–0.76	0.0016–0.027	0.0003–0.0015
Timothy	0.92–4.6	0.18–0.53	1.55–3.4	0.06–0.31	0.01–0.29	0.0025–0.019	0.0004–0.0009
Cocksfoot	1.36–4.6	0.15–0.59	1.72–4.0	0.09–0.34	0.10–0.77	0.0033–0.056	0.0004–0.0015
Meadow fescue	2.51–4.27	0.19–0.63	1.51–4.41	0.14–0.31	0.03–0.21	0.0022–0.017	0.0005–0.0023
White clover	2.66–5.3	0.19–0.47	1.54–3.8	0.15–0.29	0.05–0.48	0.0047–0.0087	0.0005–0.0001
Lucerne	2.06–5.19	0.14–0.66	1.06–3.92	0.11–0.64	0.04–0.63	0.0029–0.0073	0.0006–0.0012

TABLE 10.10 Support energy inputs to a three-year perennial ryegrass
ley receiving 325 kg N/ha/year (after Walsingham, 1976)

	MJ ha^{-1}	MJ ha^{-1} yr^{-1}
Sward establishment:		
Ploughing	1 680	
Seedbed cultivation	1 758	
Loading/carting/applying fertilizer	98	
Seed drilling	598	
Rolling	225	
Total	4 359	1 453
Fertilizer manufacture		27 931
Fertilizer application		832
Harvesting (four cuts)		10 500

efficiency of support energy use is thus much higher for legumes than for
grasses receiving nitrogenous fertilizer but does not vary greatly with the
level of fertilizer nitrogen applied to the grass (Table 10.11).

TABLE 10.11   Effect of N-fertilizer use and drying
on efficiency of support energy use

Crop	N fertilizer use (kg ha^{-1} yr^{-1})	E^a	
Perennial ryegrass	217	fresh	5.5
		dried	0.98
Perennial ryegrass	417	fresh	4.9
		dried	0.95
Lucerne		fresh	38.0
		dried	1.15

$$^aE = \frac{\text{Gross energy in product}}{\text{Support energy input to farm gate}}$$

Source: Spedding and Hoxey (1978)

Sown Forage Crops

Forage crops may be grown for grazing, cutting for ensilage, for "zero-grazing" (i.e. carted to the stock) or for leaf protein extraction (green crop fractionation).

Yields per hectare are often very high (Table 10.12), sometimes without very large inputs of nitrogenous fertilizer (Table 10.13), but the total area devoted to such crops is relatively small. Where they are used to complement grazing by providing silage or to extend the normal grazing season, the efficiency of resource use is better evaluated for the animal feeding system as a whole. Clearly, a crop that can utilize land at a season when other crops do not grow, may make a contribution to overall efficiency quite out of proportion to the quantity produced per hectare.

TABLE 10.12　Yields of forage crops

Crop	Dry matter yield (tonnes $ha^{-1}yr^{-1}$)
Marrow stem kale	6.0
Rape	1.6–6.0
Rye	1.2–3.2
Forage maize	10–15
Whole crop beans	5–10
Sunflower	7–10
Sugar cane	35–40

Sources: Bregger and Kidder (1959); M.A.F.F. (1971); J.C.O. (1975).

TABLE 10.13　Fertilizer nitrogen inputs required

	Input of N fertilizer (kg ha^{-1})	Sources[a]
Marrow stem kale	50–175	(1)
Rape	10–125	(1)
Rye	0–100	(1)
Forage maize	62.5–150	(1)
	1.25	(2)
Whole-crop beans	0–25	(1)
Sunflower	50–100	(2)
Sugar cane	0–92	(3)
Grass (for grazing)	up to 525	(1)

[a]Sources: (1) M.A.F.F. (1973); (2) J.C.O. (1975); (3) Bregger and Kidder (1959)

Whole-crop cereals, including maize, beans and peas may also be used to provide large crops for silage but comparisons between these and the grain they could have yielded need to take the animal conversion system into account. Since the latter must be less efficient than a crop system by itself, in the production of nutrients for human consumption, the system involving forage crops *may* be more efficient than one based solely on grassland but is unlikely to be better than crop production for grain.

The justification for growing forage crops, therefore, is usually based on the use of a small area of arable land to improve the efficiency with which associated grassland is utilized, or it is related to the value of particular crop rotations in a predominantly arable system.

Forage crops often require a higher labour input (Table 10.14) and very uneven labour profiles (Fig. 10.3). Their use of support energy (Table 10.15) varies with the fertilizer requirement and the method of harvesting but their high yields allow an efficiency of support energy use that is comparable to that of pasture (Table 10.16).

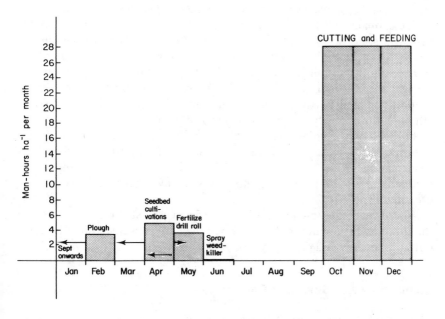

FIG. 10.3   Labour profile for growing, cutting and feeding kale (after Nix, 1976)

TABLE 10.14   Labour requirements for sown forage crops

	Man hr ha^{-1}	Sources[a]
Kale: cut	96	(1)
Kale: grazed	20	(1)
Kale (growing only)	14	(2)
Flatpoll cabbage (growing only)	89	(2)
Fodder rape (growing only)	17	(2)
Maize silage (growing and harvesting)	16	(2)

[a]Sources: (1) Nix (1976); (2) Dench and Buchanan (1977).

Vegetables and Salad Crops

Leafy vegetables, such as cabbages, Brussels sprouts and spinach, and some salad crops, such as lettuces, are grown on a farm scale. Their inputs of labour (Table 10.17) and fertilizer (Table 10.18) are relatively high but so are their yields of energy and protein per hectare (Tables 10.19 and 10.20).

The cabbage will serve as a good example of some of the special features of a farm-scale vegetable. Efficiency of use of both solar and support energy are high (Table 10.21), provided that the crop is eaten fresh. It *can* be, of course, but it is usually cooked and this influences the efficiency of support energy use considerably (Table 10.22), partly because of losses during preparation and partly because of the input of energy during cooking, although this is much less than is involved with potatoes.

Storage costs may be small, since cabbages can be left in the field until required. This feature, incidentally, is quite an important attribute for garden and allotment crops, where it also has the effect of spreading labour demand. In such situations, quite exceptionally high yields may also be obtained (Table 10.23).

Where vegetables are fed to animals, efficiency must also be calculated on the ultimate production but, even when they are consumed directly, gross output of energy and protein can be misleading.

Stems, Flowers, Bulbs and Corms

These are of relatively little importance, the exception being sugar cane which is a major world crop but which is dealt with separately in Chapter 12, and the quantities involved are small. The more important examples are given in Table 10.24.

TABLE 10.15   Use of support energy by forage crops (after Spedding *et al.*, 1979)

	Recommended nitrogen fertilizer level (kg ha^{-1})	Fertilizer	Support energy inputs (J × 10^9 ha^{-1})					Total
			Fuels		Machinery			
			Establishment	Harvest	Establishment	Harvest[a]		
Kale								
cut	100	13.1	2.18	2.66	1.44	2.64		22.02
grazed	100	13.1	2.18	2.66	1.44	—		19.38
Cabbage	150	14.9	2.18	2.22	1.44	2.20		22.94
Rape	75	4.2	2.18	1.94	1.44	1.92		11.68
Maize	90	8.6	2.13	1.94	1.56	1.92		16.15

[a]Cut and carted

TABLE 10.16   Efficiency ratios for support energy use in fodder crop systems (after Spedd et al., 1979)

	Total support energy input (J × 10⁹ ha⁻¹)	Yield kg DM ha⁻¹	$E = \dfrac{\text{Output (kg DM ha}^{-1})}{\text{Support energy (J} \times 10^9 \text{ ha}^-}$
Kale			
cut	22.02	7 300	332
grazed	19.38	6 500[a]	335
Cabbage	22.94	6 500	283
Rape	11.68	4 700	402
Maize	16.15	11 700	724
Italian ryegrass	40.8	16 900	414
Perennial ryegrass	37.4	14 600	390

[a]Estimated

TABLE 10.17   Labour requirements for leafy vegetables and salad crops

Crop	Man hr ha⁻¹	Sources[a]
Brussels sprouts	214	(1)
hand picked	346	(2)
single harvest machine picked	247	(2)
Cabbage	209	(1)
different types	185–497	(2)
Calabrese	150	(3)
Cauliflower	162	(1)
Lettuce	309	(2)
heated crops under glass	2 581–2 680	(4)

[a]Sources: (1) Nix (1976); (2) ADAS Costings (1974); (3) ADAS Costings (1976); (4) NAAS (1971).

TABLE 10.18 Fertilizer requirements
for leafy vegetables and salad crops

Crop	N (kg ha^{-1})
Brussels sprouts	187.5–375
Cabbage	62.5–362.5
Cauliflower and broccoli	112.5–312.5
Lettuce	
in open	75 –150
glasshouse	117

Source: M.A.F.F. (1973) — assuming
nutrient index for soil of 2–3

TABLE 10.19 Yields of energy and protein from leafy vegetables and salad crops

Crop		Yield (kg ha^{-1})	Fresh material Energy content (MJ kg^{-1})	Protein content (%)	Energy yield (MJ ha^{-1})	Protein yield (kg ha^{-1})
Cabbage	UK	24 900	0.84	2.0	20 916	498
	USA	26 904			22 599	538
Brussels sprouts	UK	12 000	1.09	3.5	13 080	420
	USA	11 210			12 219	392
Lettuce	UK	18 800	0.46	1.2	8 648	226
	USA	23 541			10 829	282

Sources: Mortensen and Bullard (1970); NEDO (1973)

TABLE 10.20   Possible yields of tropical leafy vegetables

Crop	Yield (DM kg ha^{-1})	Protein[a] yield (kg ha^{-1})
Amaranth *Amaranthus cruentus*	2 445–4 916	700–1 613
Malabar spinach *Basella alba*	1 812–5 316	456–1 594
Chinese mustard cabbage *Brassica chinensis*	2 479–2 649	706–969
Collards *Brassica oleracea* var. *acephala*	3 597–5 393	1 025–1 738
Jute mallow *Corchorus olitorius*	1 830–2 943	438–1 019
Swamp cabbage *Ipomoea aquatica*	3 503–4 907	1 119–1 750
Garden huckleberry *Solanum melanocerasum*	1 264–3 025	400–1 275
Waterleaf *Talinum triangulare*	1 779–3 093	425–988

[a]N × 6.25

Source: Schmidt (1971)

They have a wide variety of uses from conventional vegetables and salads to flavouring and as delicacies. Productivity per unit area of land can be quite high (Table 10.25). The category is a very mixed one: whilst asparagus and bamboo shoots are true stems, it is the petioles of both rhubarb and seakale that are eaten and in kholrabi the stem is specially modified as a storage organ.

With globe artichokes, cauliflower annd broccoli, the flower head or shoot is consumed.

Few of the bulbs, which have a stem core surrounded by storage leaves, are eaten, except for the onion family (*Allium* spp.): garlic, leeks, shallots and chives are used for their flavour and, sometimes, for their anthelmintic properties. They are rich in sugar and pungent allyl sulphides and impart flavour out of proportion to the quantities used. Thus they may be considered as efficient sources of flavour and any medicinal properties may also add value quite unrelated to their nutrient content.

TABLE 10.21 Efficiency of solar and support energy use in cabbage production

	Yield of cabbage[a] (t fresh weight per ha)	Support[b] energy used (MJ ha⁻¹)	Solar[c] energy available (J × 10¹² ha⁻¹)	$E = \dfrac{(kg)\ cabbage}{(MJ)\ support\ energy}$	$E = \dfrac{(kg)\ cabbage}{(J \times 10^{12})\ solar\ radiation}$
Spring cabbage	15.6	17 593	18.8	0.89	0.83
Early summer cabbage	23.2	44 177	23.4	0.52	0.99
Late summer/ autumn cabbage	28.5	33 593	26.4	0.85	1.78
Winter cabbage	24.0	38 129	29.6	0.63	0.81

[a]M.A.F.F. (1967) Average yields 1952–65
[b]Shiels (1978)
[c]Assuming average planting and harvesting dates; solar radiation data from de Jong (1973)

TABLE 10.22 Effect of cooking cabbage on efficiency of support energy use

	Fresh cabbage	Cooked cabbage	
Energy input for production[a]	30 710	30 710	MJ ha^{-1} yr^{-1}
Yield of cabbage for eating	23 600	20 060	kg ha^{-1}
Energy cost of cooking cabbage[b] at 1.58 MJ/kg	—	31 695	MJ ha^{-1}
Total support energy input	30 710	62 405	MJ ha^{-1} yr^{-1}
Support energy used per kg cabbage	1.30	3.11	MJ

[a]Shiels (1978)
[b]Irving (1978). (The value for potatoes is 3.55 MJ kg^{-1})

The only corm of any importance is the taro or cocoyam. It is rich in starch and eaten in much the same way as potatoes. The leaves and petioles can also be utilized as vegetables; it therefore has good potential as a food crop in the tropics.

Other Vegetables

Maize, peas and beans as grain and pulse crops have been dealt with in Chapters 6 and 7. However, eaten at an earlier stage these crops are useful vegetables.

Corn-on-the-cob or sweet corn may be a specially produced crop from varieties having the particular attribute required, of more sugars in the kernel than the starchy grain varieties, or simply the grain varieties used before they are fully mature.

The same is true for many varieties of peas and beans, but, particularly in the case of beans, the pod is also consumed when eaten as a green vegetable.

This may have the effect of altering the yield and composition of edible material per unit of land and/or extending the period in the year when useful food is supplied, which could be important in subsistence farming.

Trees and Shrubs

Much less information is available on the productivity of browse trees and shrubs (Burrows and Beale, 1970; McKell, 1975; Rajalakshmi and Rama-krishman, 1969; Skerman, 1977) and, where they are used on a substantial scale, they tend to be multipurpose or the only plants that the environment will support.

TABLE 10.23   Yields of vegetables (kg ha^{-1})

Crop	Garden[a] UK	Farm scale[b] UK	Horticultural[c] USA	Commercial[d] UK
Beans				
broad	40 804[e]			10 900
dwarf french	32 374	11 250	7 847 (Snap beans)	7 500
runner	16 254			7 500
Beetroot	65 780		26 904 (beets)	29 600
Broccoli, sprouting	32 374			
Brussel sprouts	16 277	15 000	11 210	11 800
Carrots	65 780	35 000	37 834	38 700
Kale	32 374			
Leeks	49 335			22 800
Onions	49 335	32 500	39 235	28 100
Parsnips	43 405			19 400
Peas	16 322[e]		5 045	7 700
Potatoes				
early	54 256	{ 18 000 / 25 000 }		
maincrop	39 213	36 000	26 904	
Shallots	95 745			
Spinach	49 335		14 013	
Swedes and turnips	65 780		22 420	
Tomatoes, outdoors	32 554			32 700

[a]Which, Feb. 1975; [b]Nix (1976); [c]Mortensen and Bullard (1970); [d]M.A.F.F. (1977)

[e]Including pod

TABLE 10.24   The major examples of stem, bulb, flower and corm crops

Asparagus	*Asparagus officinalis*
Bamboo	*Bambusa vulgaris*
Rhubarb	*Rheum rhaponticum*
Seakale	*Crambe maritima*
Kohlrabi	*Brassica oleracea* var. *gongyloides*
Celery	*Apium graveolens*
Globe artichoke	*Cynara scolymus*
Cauliflower	*Brassica oleracea* var. *botrytis*
Broccoli	*Brassica oleracea italica*
Onion	*Allium cepa*
Shallot	*Allium ascalonicum*
Garlic	*Allium sativum*
Leek	*Allium porrum*
Taro or cocoyam	*Colocasia esculenta*

TABLE 10.25   Productivity of some of the crops listed in Table 10.24, as reported in the literature

Crop	Yield (kg ha^{-1} yr^{-1})	Fresh material Energy content (MJ kg^{-1})	Protein content (%)	Energy yield (MJ ha^{-1})	Protein yield (kg ha^{-1})
Asparagus	6 726	1.13	2.8	7 600	188
Celery	100 890	0.3	0.8	30 267	807
Onion	39 235	1.6	1.5	62 776	588
Cauliflower	24 886	0.92	2.2	22 895	547
Taro (cocoyam)	75 000	3.6	1.5	270 000	1 125

Sources: Chatfield (1954); Mortensen and Bullard (1970); Kay (1973)

Thus acacias in Africa can fix their own nitrogen and tap enormous depths of soil for water. Sagebrush in the USA and saltbush in Australia also withstand very dry conditions: these and some other species are also able to grow in soils with relatively high salinity.

Multipurpose trees may be grown to provide shade for grazing animals and perhaps fruits as well. The cashew tree (*Anacardium occidentale*) is an African example of this.

Acacias in Africa are often too tall to be browsed but stock eat the shed seeds, pods and leaves. Wattle (*Acacia mollissima*) is an extremely palatable tree, as are *Leucaena glauca* and *L. leucocephala*, which have to be protected for the first year to avoid destruction by cattle.

References

A.D.A.S. (1974). Costings.

A.D.A.S. (1976). Costings.

Black, J. N. (1964). An analysis of the potential production of swards of subterranean clover (*Trifolium subterranean* L.) at Adelaide, South Australia. *J. appl. Ecol.* 1, 3–18.

Bregger, T. and Kidder, R. W. (1959). Growing sugarcane for forage. Circular S–117 University of Florida Agricultural Experiment Stations, Gainsville, Florida.

Burrows, W. H. and Beale, I. F. (1970). Dimension and production relations of mulga (*Acacia aneura,* F. Muell.) trees in semi-arid Queensland. Proc. XI Int. Grassld Congr., 33–35. Surfers Paradise 1970.

Chatfield, C. (1954). Nutr. Stud. No. 11 Food Composition Tables. F.A.O., Rome.

Coleman, R. L., Holder, J. M. and Swain, F. G. (1966). Production from dairy cattle on improved pasture in a subtropical environment. Proc. Xth Grassland Congress, Helsinki, July. 499–503.

Dench, J. A. L. and Buchanan, W. I. (1977). Fodder crops. Agricultural Enterprise Studies in England and Wales. Economic Report No. 50. University of Reading Department of Agricultural Economics and Management.

Downes, R. W. (1969). Differences in transpiration rates between tropical and temperate grasses under controlled conditions. *Planta (Berl.)* 88, 261–273.

F.A.O. (1971). Agricultural commodity projections 1970–1980. Vol. 1. F.A.O., Rome.

F.A.O. (1974). Production Yearbook 1973, 27 F.A.O., Rome.

Gilchrist, E. C. (1967). A place for Stylo in New Queensland pastures. *Queensland Agric. J.* 93, 344–9.

Green, J. O. and Corrall, A. J. (1974). Evaluation of forage crops. *In* "Silver Jubilee Report 1949–74", (Eds C. R. W. Spedding and R. D. Williams), Grassland Research Institute, Hurley. 27–33.

Hilldreth, A. C., Magness, J. R. and Mitchell, J. W. (1941). Effects of climatic factors on growing plants. Yearbook of Agriculture, 1941. U.S.D.A., Washington, D.C.

Irving, R. J. (1978). Dept. of Home Economics, University of Surrey. Personal communication.

JCO (Joint Consultative Organisation for Research and Development in Agriculture and Food) (1975). Arable crops and forage board: Grassland and forage committee. Report of the working group on forage crops (other than grasses and legumes).

de Jong, B. (1973). "Net Radiation Received by a Horizontal Surface at the Earth", Delft University Press, Delft.

Kay, D. E. (1973). Root Crops. Tropical Products Institute Crop and Product Digest No. 2. T.P.I., London.

Le Du, Y. (1976). Grassland Research Institute, Hurley. Personal communication.

M.A.F.F. (1967). "Horticulture in Britain. Part I, Vegetables", H.M.S.O. London.

M.A.F.F. (1971). Nutrient allowances and composition of feeding stuffs for ruminants. A.D.A.S. Advisory Paper No. 11.

M.A.F.F. (1973). Fertilizer Recommendations. Bull. 209. H.M.S.O., London.

M.A.F.F. (1977). Output and Utilization of Farm Produce in U.K. 1975/6.

Maximov, N. A. (1929). "The Plant in Relation to Water", George Allen and Unwin, London.

McBratney, J. and Laidlaw, A. S. (1974). The suitability of various grasses, cutting intervals and nitrogen levels for conservation systems as judged by dry matter production. 47th Annual Report Agricultural Research Institute of N. Ireland. 1973–74.

McKell, C. M. (1975). Shrubs — a neglected resource of arid lands. In "Food: Politics, Economics, Nutrition and Research", (Ed. P. H. Abelson), American Association for the Advancement of Science, Washington, D.C.

Mortensen, E. and Bullard, E. T. (1970). "Handbook of Tropical and Sub-Tropical Horticulture", Dept. of State, Agency for Int. Dev. Washington D.C. 20523, U.S.A.

N.A.A.S., (1971). Horticultural Enterprises Booklet. 2, "Glasshouse Lettuce."

N.E.D.O. (1973). "U.K. Farming and the Common Market: Outdoor Vegetables".

Nix, J. (1976). "Farm Management Pocketbook", 7th edn, Wye College, Kent.

Rajalakshmi, R. and Ramakrishnan, C. V. (1969). Horticulture in relation to nutritional requirements. Pl. Fds Hum. Nutr. 1, 79–92.

Schmidt, D. R. (1971). Comparative yields and composition of eight tropical leafy vegetables grown at two soil fertility levels. Agronomy J. 63, 546–550.

Shiels, L. (1978) University of Reading, personal communication.

Skerman, P. J. (1977). "Tropical Forage Legumes". F.A.O. Plant Production and Protection Series No. 2. F.A.O., Rome.

Shepherd, F. W. (1975). Vegetables. In "Food Protein Sources" (Ed. N. W. Pirie), IPB 4. Cambridge University Press, Cambridge.

Spedding, C. R. W., Bather, D. M. and Shiels, L. A. (1979). An assessment of the potential of U.K. agriculture for producing plant material for use as an energy source. Report to Department of Energy (Energy Technology Support Unit) University of Reading. August.

Spedding, C. R. W. and Diekmahns, E. C. (1972). "Grasses and Legumes in British Agriculture". C. A. B. Farnham Royal, England.

Spedding, C. R. W. and Hoxey, A. M. (1978). The biological efficiency of grassland farming. In Pasture and Forage Production in Seasonally Arid Climates. Proc. 6th General Meeting of the European Grassland Federation, Madrid 1975. 183–195.

Tayler, J. C. (1970). Dried forages and beef production. J. Br. Grassld Soc. 25(2), 180–90.

Vincente-Chandler, J., Silva, S. and Figarella, J. (1959). The effect of nitrogen fertilization and frequency of cutting on the yield and composition of three tropical grasses. Agron. J. 51(4), 202–6.

Walsingham, J. M. (1976). Energy for grass. Ann. Report Grassland Research Institute 1975, Hurley. 138–144.

Waslingham, J. M., Large, R. V. and Newton, J. E. (1975). The effect of time of weaning on the biological efficiency of meat production in sheep. Anim. Prod. 20, 233–41.

Which (1975). Consumers Assoc. "Handyman Which", Feb. 1975.

Whitney, A. S., Kanehiro, Y. and Sherman, G. D. (1967). Nitrogen relationships of three tropical forage legumes in pure stands and in grass mixtures. Agron. J. 59, 47–50.

11

FRUIT AND NUT CROPS

Fruit Crops

Although very attractive and widely consumed for a very long time, fruit crops are not major suppliers of energy. They supply vitamins (notably C) and minerals, however, and they are often of local, if seasonal, importance. They tend to be watery and difficult to store unless dried, cooked in sugary syrups, fumigated, canned or frozen; in fact, most are preferred raw.

The most important fruit crops are listed in Table 11.1, with an indication of their chemical composition. A great many more species are of limited

TABLE 11.1   The most important[a] fruit crops and their chemical composition

Fruit	Composition[b] per 100 g edible proportion of fresh uncooked fruit					
	Water (g)	Protein (g)	Fat (g)	Carbohydrate (g)	Vitamin A (I.U.)	Ascorbic acid (mg)
Grape	81.9	1.4	1.4	14.9	80	4
Citrus (e.g. Orange)	87.2	0.9	0.2	11.2	(190)	49
Banana	74.8	1.2	0.2	23.0	430	10
Tomato	94.1	1.0	0.3	4.0	1 100	23
Apple	84.1	0.3	0.4	14.9	90	5
Mango	81.4	0.7	0.2	17.2	6 350	41
Pear	82.7	0.7	0.4	15.8	20	4
Peach	86.9	0.5	0.1	12.0	880	8
Plum	85.7	0.7	0.2	12.9	350	5
Pineapple	85.3	0.4	0.2	13.7	130	24

[a]after FAO (1974)
[b]after Spector (1956)
Values based on inadequate evidence are enclosed in parentheses

143

commercial importance, such as lemons, breadfruit, figs, cherries, avocados, and strawberries, and even more are of only local significance. Citrus fruits dominate the subtropics, mangos the tropics and apples (and pears) predominate in temperate regions (Borgstrom, 1973).

Yields of energy and protein per unit of land (Table 11.2) can be quite high but the requirement for labour (Table 11.3) and support energy (Table

TABLE 11.2   Yields of energy and protein per unit of land from fruit crops

Crop	Yield (t ha^{-1})	Energy (GJ ha^{-1})	Protein (kg ha^{-1})	Location
Grape	19.1	55.9	267	Cyprus[a]
Citrus (e.g. orange)	56.5	106.4	509	Cyprus[a]
Banana	40.0	147.0	480	Central America[c]
Tomato (under glass)	1 29.5	108.8	1295	UK[b]
Apple	11.7	28.3	35	UK[b]
Mango	8.1	22.4	57	South Africa[d]
Pear	5.5	14.4	38	UK[b]
Peach	33.4	64.2	167	Cyprus[a]
Plum	3.7	7.7	27	UK[b]
Pineapple	4.6	10.1	19	Taiwan[e]

[a]Papachristodoulou (1970)
[b]M.A.F.F. (1977)
[c]Purseglove (1972)
[d]Singh (1968)
[e]Collins (1968)

N.B. Energy and protein contents are calculated from Spector (1956) as though entire crop edible. Wastage will be greatest in citrus, bananas, mango and pineapple.

TABLE 11.3   Examples of the labour requirements (man hours) of fruit crops

Crop	Situation	h ha^{-1} yr^{-1}
Grapes	Seedless grown in Cyprus[a]	1 238
Oranges	Valencia grown in Cyprus[a]	1 653
Apples	Dessert grown in UK[b]	318
Pears	Dessert grown in UK[b]	190
Peaches	Grown on Cyprus plains[a]	1 880
Plums	Grown on Cyprus plains[a]	1 865

[a]Papachristodoulou (1970)
[b]Calculated from M.A.F.F. (1978)

11.4) are also high. The total world contribution of fruit and nuts to man's dietary energy supply is less than 10% (Brown and Finsterbusch (1972) estimate 9.6%, including vegetables). Since fruits are chiefly valued for their flavour, the quality of the product is very important, but, paradoxically, some of the best-flavoured fruits are virtually excluded from international trade because they are also too delicate to withstand the hazards of transport (Janick et al., 1969). This applies even to some of the most delicious varieties within a crop species (e.g. peaches).

TABLE 11.4 Support energy used in fruit production in the UK (after Shiels, 1978)

		$GJ\ ha^{-1}\ yr^{-1}$
Soft fruit	Blackcurrants	23
	Raspberries	23
	Strawberries	12
Top fruit	Apples	
	cooking	19
	dessert	22
	Plums	18
Vines	Grapes (Guyot system)	32

There is relatively little point in comparing the efficiency of one fruit crop with another, since the desire for an orange cannot be met by a pineapple, whatever the relative efficiencies of the two crops. There is a great deal of point, however, in comparing varieties within a species and in comparing different methods of production for such highly valued products. Of course, wherever a fruit is a major supplier of dietary nutrients, choice of species may well be important and based on efficiency of land and labour use.

Figure 11.1 illustrates the components of efficiency for fruit crops, as exemplified by the apple. The yield of apples per ha is determined by yield per tree, the number of trees, the number of fruit per tree and the influence of pests and disease. In general, higher inputs of labour, machinery and biocides tend to increase yield but the efficiency with which some resources are used may decline (Table 11.5).

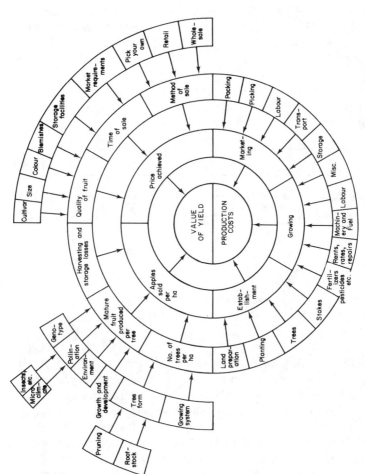

FIG. 11.1   Components of economic efficiency for fruit crops, as exemplified by apple production (per ha). Only the more important components are shown: the diagram can be extended as required. The factors affecting "growth and development" of the fruit will be similar to those on the two outer circles of Fig. 6.1. Only the direct connections between the circles have been shown for the sake of simplicity.

TABLE 11.5   Support energy inputs (MJ ha⁻¹) to apple production

	Cider apples	Dessert apples
Support energy input to:		
Fertilizer	3 703	12 239
Herbicides	592	1 344
Insecticides and fungicides	608	1 215
Tractor fuel	3 110[a]	3 110
Repairs and depreciation	1 874[a]	1 874
Irrigation	—	6 935
Total (MJ ha⁻¹)	9 887	26 717
Yield (t ha⁻¹)	20.0	40.0
Support energy input (MJ kg⁻¹ apples)	0.49	0.67

[a]In the absence of figures it is assumed that these are the same as for dessert apples

"Vegetable" Fruits

Most of the fruits so far mentioned are grown as trees or shrubs but there are also many fruits that are classed as vegetables. The most important ones are herbaceous annuals and the main botanical families are the Solanaceae and the Cucurbitaceae. The first includes the tomato and the eggplant; the second is the gourd family, including the melons, squashes and cucumbers. Both include sprawling and trailing plants and have generally involved much labour, although mechanization is increasing. Although most of the vegetable fruits are even less important commercially than the other fruit crops, the gourds are main items of the diet of some primitive, tropical societies (Janick *et al.*, 1969) and the tomato is a major world food plant. Tomato production totals some 39 million metric tonnes annually, mostly in Europe and North America, occurring both as an outdoor and a glasshouse crop.

The productivity of the tomato can be very high (see Table 11.6 for yields of energy and protein per ha) but labour demands are also substantial (Table 11.7). Efficiency of fertilizer use (Table 11.8) is related to fertilizer inputs (Fig. 11.2).

Support energy inputs are very large for glasshouse tomato crops, most of the energy being used for heating (Table 11.9). Clearly, more effective glasshouse design could have an enormous influence on this aspect of efficiency and the increasing cost of oil will presumably bring many changes in this direction.

TABLE 11.6   The productivity of tomatoes

Situation	Yield (kg ha^{-1})	Energy yield (GJ ha^{-1})	Protein yield (kg ha^{-1})	Sources[a]
Kenya — field grown	17 566	10.4	158	(1)
USA — field grown (average)	9 240	5.5	83	(2)
Haiti — field grown (maximum)	25 850	15.3	233	(2)
Cyprus — field grown	28 647	16.9	258	(3)
UK — premium glasshouse yields from				
mid-February planting	184 800	109.0	1 663	(4)
mid-March planting	160 160	94.5	1 441	(4)
mid-April planting	135 520	79.9	1 220	(4)
UK — average glasshouse yields	112 000	66.1	1 008	(5)

[a]Sources: (1) Kenya Department of Agriculture (1966); (2) King (1971); (3) Papachristodoulou (1970); (4) Kingham (1973); (5) M.A.F.F. (1977)

TABLE 11.7   Examples of labour inputs (man h ha^{-1}) to tomato production

Operation	Field-grown[a] tomatoes in Cyprus — planted in April	Operation	Glasshouse[b] tomatoes in UK — planted in February
Ploughing	11	Forking	741
Opening furrows	8	Propagation	1 383
Transplanting	53	Planting	494
Hoeing	120		
Earthing up	90		
Pruning and staking	181	Twisting and	7 410
Fertilizing	30	sideshooting	
Irrigation	361		
Plant protection (spray)	75	Spraying	124
Harvesting	1 128	Picking	2 779
Miscellaneous	165	Grading and packing	1 853
		Clearing crop	371
Total labour (h ha^{-1})	2 222	Total	15 155
Yield of tomatoes (kg ha^{-1})	28 647		252 933
Labour hours per kg tomatoes	0.078		0.06

[a]Papachristodoulou (1970)
[b]M.A.F.F. (1962)

TABLE 11.8   Output of field-grown tomatoes (cultivar Floradel) per unit of fertilizer in Puerto Rico (after Abrams *et al.*, 1975)

Fertilizer levels (kg ha⁻¹)			Marketable yield		Total yield	
			kg protein	MJ per	kg protein	MJ per
N	P	K	per kg N	kg N	per kg N	kg N
112.1	224.2	224.2	1.48	97	1.76	115
224.2	224.2	224.2	0.81	53	0.94	61
448.4	224.2	224.2	0.47	31	0.53	35

TABLE 11.9   Support energy inputs (GJ ha⁻¹) to glasshouse tomato production in the UK (after Shiels, 1978)

	Early tomatoes Heated crop Planted February	Late tomatoes Unheated Planted April
Support energy input to:		
Fertilizers	189.9	165.3
Insecticides and fungicides	21.3	21.3
Irrigation	63.7	51.9
Machinery (fuel)	4.1	4.0
Machinery (depreciation and repair)	1.3	1.3
Heating oil	26 562	—
Fuel for soil sterilization	9 531	9 531
Electricity	960	—
Propane (for CO_2 enrichment)	971	—
Support strings	16.2	16.2
Glasshouse depreciation and repair	1 729	1 153
Total energy inputs (GJ ha⁻¹)	40 050	10 944
(GJ kg⁻¹ tomatoes)	0.16	0.09

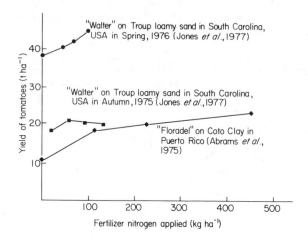

FIG. 11.2   Tomato yield responses to nitrogen application

Nut Crops

It is surprising that nuts make so little contribution to the World's food supply. By comparison with the seeds of grasses, they appear a more promising basis for food production; yet they have been relatively little used and very little cultivated. Indeed, it is considered that nuts are still more often gathered from the wild than most other foods. Wild trees are the source of most Brazil nuts, cashews, black walnuts, certain hickory nuts and pecans (Janick *et al.*, 1969). Nuts *are* cultivated, of course, such as almond, filbert, cashew and domesticated pecans, but the scope for further development on certain classes of land seems considerable.

The advantages are their high nutritional value and their good keeping qualities. They are rich in oil and fat and contain moderate amounts of carbohydrate and protein (Table 11.10) and their low moisture content makes them less costly to transport than fruit.

Where nut crops are cultivated in orchards, their labour requirements are high (Table 11.11) but their yields per unit of land (Table 11.12) and per unit of support energy (Table 11.13) may also be very high. However, relatively little improvement by breeding has yet taken place in the nut-producing species. One major disadvantage is the long time needed to establish a fruit- or nut-bearing tree (Table 11.14) and the latter are often very slow-growing: they do have a long productive life, once established, and the pistachio is one of the few dicotyledonous trees known to last for over a thousand years.

Since many of these trees are best grown some distance apart, intercropping with pasture may improve overall yield.

TABLE 11.10   Composition of nuts (edible material as purchased)

	Water (g per 100 g)	Energy (kJ g^{-1})	Protein (g per 100 g)	Fat (g per 100 g)	Carbohydrate[a] (g per 100 g)	Sources[b]
Almonds	4.7	23	16.9	53.5	4.3	(1)
Prunus amygdalus	4.7	25	18.6	54.2	19.5	(2)
			18.6	54.2	19.5	(3)
			21.0	61.0	13.0	(4)
Brazil nuts	8.5	25	12.0	61.5	4.1	(1)
Bertholletia excelsa	4.6	29	14.0	66.9	10.9	(2)
			14.3	66.9	10.9	(3)
			14.0	68.0	9.0	(4)
Cashew nuts	5.2	25	17.2	45.7	29.3	(2)
Anacardium occidentale			17.2	45.7	29.3	(3)
			21.0	46.0	25.0	(4)
Coconut (meat)	42.0	14	3.2	36.0	3.7	(1)
Cocos nucifera	48.0	15	4.2	34.0	12.8	(2)
			3.5	35.0	9.4	(3)
Hazelnuts or Filberts	41.0	16	7.6	36.0	6.8	(1)
Corylus spp.	6.0	26	12.7	60.9	18.0	(2)
			12.8	62.0	17.0	(3)
Walnut	23.5	22	10.6	51.5	5.0	(1)
Juglans spp.	3.5	27	14.8	64.0	15.8	(2)
			16.0	64.0	15.5	(3)

[a]Some figures include fibre in the carbohydrate total
[b]Sources: (1) Paul and Southgate (1978); (2) Deatherage (1975); (3) Douglas and Hart (1976); (4) Fisher and Bender (1975)

N.B. A number of sources have been used to illustrate the variations that occur

Harvesting is less of a problem with nuts than with fruit, since there are no real disadvantages in allowing them to fall to the ground where they can simply be collected.

TABLE 11.11   Labour requirements of nut production in orchards

	Man h ha^{-1} yr^{-1}	Yield kg ha^{-1} yr^{-1}
Almonds	2 558	11 250

Source: Papachristodoulou (1970)

TABLE 11.12   Nut yields per unit of land

Species	Annual yield (kg ha^{-1})	Sources[a]
Chestnuts (USA and France) *Castanea* spp.	17 600–28 000	(1)
Walnuts	25 000–38 000	(1)
Walnuts (maximum yield)	9 000	(2)
Pecans (N. America) *Carya* spp.	23 000–28 000	(1)
Pecans (Australia)	8 400	(3)
Hazelnuts	23 000–30 000	(1)
Hazelnuts	1 256	(4)
Filbert (maximum yield)	4 500	(2)
Filbert (exceptional yield)	4 500	(3)
Pistachio	1 000–2 000	(3)
Almonds	1 256	(4)
Almonds	11 250	(5)

[a]Sources: (1) Douglas and Hart (1976); (2) Westwood (1978); (3) Woodruff (1979); (4) Janick et al. (1969); (5) Papachristodoulou (1970)

N.B. References do not state if yield weights include shells or not, or are for fresh or dry weight

TABLE 11.13   Yield of nuts (kg) per unit of support energy (GJ) (after Cervinka et al., 1974)

Almonds	34
Walnuts	22

TABLE 11.14   Time required to establish fruit and nut trees

	No. of years to first bearing	No. of years to full production
Apple	5	10
Orange	5	10–12
Plum	6	
Grape	3	4–5
Banana	1–2	
Mango	4–5	10–20
Filberts and Cobnuts	3–4	6
Walnuts		
grafted	7	
from seed	10–15	
Pistachio	5	10–20
Cashew	3	7
Almond	7	

Sources include: Hyams (1975); Janick et al. (1969); Papachristodoulou (1970); Buckley (1978); Purseglove (1968)

References

Abrams, R., Cruz-Pérez, L., Pietri-Oms, R., and Julia, F. J. (1975). Effect of fertilizer N, P, K, Ca, Mg and Si on tomato yields in an oxisol. *J. Agriculture of the University of Puerto Rico* **59**, 26–34.

Borgstrom, G. (1973). "World Food Resources", Intertext Books, Buckinghamshire.

Brown, L. R. and Finsterbusch, G. W. (1972). "Man and His Environment: Food". Harper and Row, New York.

Buckley, W. R. (1978). University of Reading. Personal communication.

Cervinka, V., Chancellor, W. J., Coffett, R. J., Curley, R. G. and Dobie, J. B. (1974). "Energy requirements for Agriculture in California". University of California, and California Deparment of Food and Agriculture.

Collins, J. L. (1968). "The Pineapple", World Crop Series. Leonard Hill Books, London.
Deatherage, F. E. (1975). "Food for Life", Plenum Press, New York.
Douglas, J. Sholto and Hart, R. A. de J. (1976). "Forest Farming", Watkins, London.
F.A.O. (1974). Production Yearbook 1973. Vol. 27. F.A.O., Rome.
Fisher, P. and Bender, A. (1975). "The Value of Food", 2nd edn, Oxford University Press, Oxford.
Hyams, E. (1975). "Survival Gardening", John Murray, London.
Janick, J., Schery, R. W., Woods, F. W. and Ruttan, V. W. (1969). "Plant Science", W. H. Freeman, San Francisco.
Jones, T. L., Jones, U. S. and Ezell, D. O. (1977). Effect of nitrogen and plastic mulch on properties of Troup loamy sand and on yield of "Walter" tomatoes. *J. Amer. Soc. Hort. Sci.* **102(3)**, 273–275.
Kenya Department of Agriculture (1966). Horticultural Section. Horticultural Handbook. Vol. 1.
King, K. W. (1971). The place of vegetables in meeting the food needs in emerging nations. *Economic Botany* **25(1)**, 6–11.
Kingham, H. G. (ed.) (1973). "The UK Tomato Manual". Grower Books, London.
M.A.F.F. (1962). Horticulture as a business: Management Handbook — Horticulture.
M.A.F.F. (1977). Output and utilization of farm produce in the U.K. 1969/70 to 1975/6. Central Statistical Office. H.M.S.O., London.
M.A.F.F. (1978). A quick guide to top fruit costings, 1978. M.A.F.F., Cambridge.
Papachristodoulou, S. (1970). Norm input-output data of the main crops of Cyprus. Cyprus Agric. Res. Inst., Ministry of Agric. and Natural Resources. Miscellaneous Publications No. 7, Nicosia.
Paul, A. A. and Southgate, D. A. T. (1978). "The Composition of Foods". 4th edn. M.R.C. Special Report No. 297. H.M.S.O., London.
Purseglove, J. W. (1968). "Tropical Crops. Dicotyledons, I", Longmans, London.
Purseglove, J. W. (1972). "Tropical Crops. Monocotyledons, 2", Longmans London.
Shiels, L. A. (1978). Personal communication.
Singh, L. B. (1968). "The Mango", World Crop Series, Leonard Hill Books, London.
Spector, W. S. (Ed.) (1956). "Handbook of Biological Data", W. B. Saunders.
Westwood, M. N. (1978). "Temperate-zone Pomology", W. H. Freeman, San Francisco.
Woodruff, J. G. (1979). "Tree Nuts", A.V.I., Westport, Connecticut.

12

SUGAR PRODUCTION

Sugary substances are produced in a variety of ways, including the tapping of maple trees in Canada to produce syrup, but the two major crops are the sugar cane (*Saccharum officinarum*) and the sugar beet (*Beta vulgaris*). Their relative importance may be judged from the quantities of sugar they contribute to world supplies (Table 12.1).

TABLE 12.1   Quantities of sugar contributed to world supplies by cane and beet

	Centrifugal sugar tonnes — raw value	Percentage of total production
Beet	33 731 521	39
Cane	52 781 723	61
Total	86 513 244	

Source: International Sugar Organization (1976)

These two crops are grown in totally different climatic regions and the only real reason for comparing them with each other is that they often supply the same market, so it is possible, for example, to consider the total use of resources in supplying the British household with sugar from beet grown in the UK compared with sugar derived from a source such as Mauritius. This can be illustrated most usefully in terms of the efficiency with which support energy is used (Table 12.2). However, processing may have a large effect on such an efficiency ratio: it is particularly noticeable that the effect of what many people would regard as a quite unneccessary, or even undesirable,

155

TABLE 12.2   Efficiency of support energy use in supplying household sugar (after Austin *et al.*, 1978)

| | Sugar from sugar beet | | Sugar from sugar cane | |
| | California, | | Queensland, | Transvaal, |
	UK	USA	Australia	South Africa
Support energy[a] used (MJ kg^{-1})	23.5	34.2	10.3	10.6

[a]Excluding the additional energy inputs required in each case to transport the sugar to the consumer

degree of processing may be greater than the unavoidable energy cost of transport (Austin *et al.*, 1978).

Sugar cane is a good example of a C_4 plant, normally grown in tropical or subtropical areas, where high temperatures and high light intensities allow such plants to use to the full their particular photosynthetic pathways. As a result yields per hectare and per unit of solar radiation are exceptionally high (Table 12.3) and may even reach 100 t ha^{-1} yr^{-1} and exceed 4% of total radiation, with some 40–45% of energy fixed as sugar. The rest of the energy fixed also has some use and sugar cane is one of the best examples of part of a crop being used to fuel the processing of the rest. Leaves and tops (trash) may be left in the field but after the stems have been taken to the factory, the fibrous portion (bagasse) is burnt to generate steam and hence electrical power (some 1300 kWh per 100 t of stem processed). Excess bagasse, boiler ash, filter mud and molasses are additional by-products: raw sugar represents about 10–16% of the starting material (Coombs, 1975; Paturau, 1969). Each of the by-products may be the starting point for industrial processes, resulting in paper, laminates, hardboard, solvents, beverages or antibiotics, or they may be used as fertilizer. Thus the efficiency of sugar production is one facet of a complex enterprise (see Fig. 12.1), just as oil seeds and animal production systems are the starting point for several industries.

TABLE 12.3   Yield of dry matter by sugar cane in Hawaii (after Cooper, 1970)

DM (t ha^{-1} yr^{-1})	64.1
kg DM per GJ of radiation	0.98

Sugar beet also gives rise to by-products (both the tops and the residual pulp are fed to animals, for example) but they are not of comparable importance.

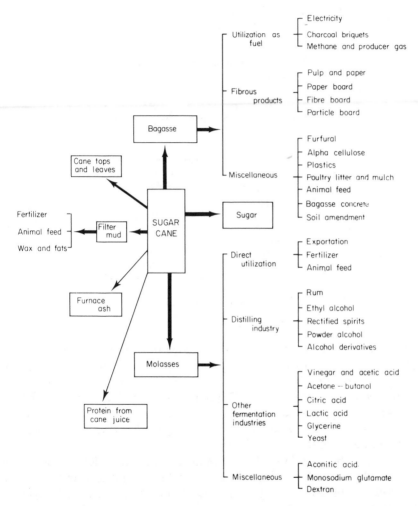

FIG. 12.1   The sugar production enterprise (from Paturau, 1969)

Sugar beet is not as productive as sugar cane per unit of land or incident solar radiation (Table 12.4) and, as a production system, it is more expensive of support energy (Table 12.5) than traditional cane production. Both crops require a lot of labour (Table 12.6).

TABLE 12.4   Yield of dry matter
by sugar beet in Holland

DM (t ha^{-1} yr^{-1})	18
kg DM per GJ of radiation	0.48

Sources: Hudson   (1975);   Cooper
(1970).

TABLE 12.6   Labour requirements for
sugar crops

	hr ha^{-1} yr^{-1}
Sugar cane	
Hawaii, 1970	107
Sugar beet	
California, 1970	70
UK	80

Sources: Heichel (1973); Nix (1976)

It is very instructive to observe how crops that are thought of as being grown primarily for their energy content, may nevertheless require as much or even more support energy, for processing, refining, transporting and distribution, than they contain. The situation becomes even more remarkable where the sugar is used primarily for sweetening purposes, rather than as a source of energy, by people whose energy intake is, on average, well in excess of their requirements.

Although the protein content of these crops is rather low (Table 12.7) total yields per ha may nevertheless be high but all of it is diverted into the by-products. In sugar cane, for example, about 56% of the total nitrogen in the plant finishes up in the bagasse.

Sugar beet is an annual crop (although a biennial species), whereas sugar cane can be harvested for several years in succession from a single planting.

Sugar cane is therefore one of the most efficient, if not *the* most efficient agricultural crop in terms of energy production, in regions where it can be grown. It has been estimated that its theoretical potential yield per hectare is about double (i.e. about 200 t ha^{-1} yr^{-1}) even the top of the range

TABLE 12.5 Support energy (GJ ha^{-1} yr^{-1}) used in producing refined sucrose from sugar beet and sugar cane (after Austin et al., 1978)

	Sugar Beet		Sugar Cane	
Support energy inputs	UK	California, USA	Queensland, Australia	Transvaal, South Africa
Fuel used in cultivations	7.1	8.9	6.7	5.1
Energy equivalent of farm machinery	5.3	8.6	9.4	9.6
Fertilizers	15.4	14.1	10.7	11.0
Irrigation (pumping)	0.0	10.2	8.9	15.2
Plant protection chemicals	1.0	1.6	0.1	0.6
Transport to factory	5.5	8.5	9.7	5.9
Energy equivalent of factory machinery	5.6	9.0	12.5	9.7
Fuel used in factory			4.1	0.0
Fuel used for refining	82.4	171.5	70.0	57.2
Total (GJ ha^{-1} yr^{-1})	122.3	232.4	132.1	114.3
Dry matter yields (t ha^{-1} yr^{-1})	8.6 (Roots) 3.4 (Tops)	16.5	43.8	42.0
Dry matter yield (t ha^{-1} yr^{-1}) per GJ of support energy	0.10	0.07	0.33	0.37

TABLE 12.7   Protein production by sugar cane and sugar beet crops

Crop	Yield DM (kg ha^{-1} yr^{-1})			Crude protein (CP) (%)	CP yield (kg ha^{-1})
Sugar cane					
West Indies	35 000			3.68	1 288
Hawaii	90 000				3 312
Sugar beet					
UK	9 000	Root	6 300^a	4.8	
		Leaves	2 700	12.5	640
Holland	18 000	Root	12 600^a	4.8	1 280
		Leaves	5 400	12.5	

^{a}Assuming the roots to be 70% of the total DM

Sources: Hudson (1975); Chapman *et al.* (1964); Scott (1977); M.A.F.F. (1975)

normally achieved in practice. This appears so promising as to justify considerable effort, by research and development, to eliminate the usual practical constraints of pests and diseases, water and fertilizer shortage, soil and climatic conditions.

It is therefore worth examining the effect of yield, in both sugar cane and sugar beet, on the efficiency with which the major resources are used in energy production. This is shown for solar radiation in Fig. 12.2, for support energy in Fig. 12.3 and for labour in Fig. 12.4. It is considerations of this kind that influence both the probability that theoretical potentials will be achieved in practice and, indeed, the desirability that they should.

Nectar

Honey production can be an important source of sugar (see Chapter 24) and is, of course, based on nectar, secreted by the nectaries of flowers of higher plants. Quantitatively, outputs per unit of land are not of the same order as crops like cane and beet, but they are not negligible. Over 500 kg of honey per hectare could be produced from the nectar of sage, thyme or maple, for example (Crane, 1975).

Furthermore, this is usually a by-product in addition to other outputs and the plants have not been bred for nectar yield. Collection by bees would be difficult to displace but it is rather inefficient energetically: the possibility of harvesting flowers as a source of sugar has not seriously been explored.

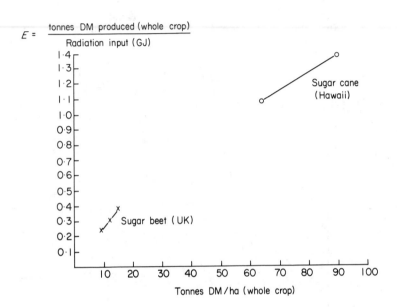

$$E = \frac{\text{tonnes DM produced (whole crop)}}{\text{Radiation input (GJ)}}$$

FIG. 12.2 Effect of yield on efficiency of use of solar radiation in the production of dry matter (DM)

$$E = \frac{\text{tonnes DM produced (whole crop)}}{\text{SE input on farm (GJ)}}$$

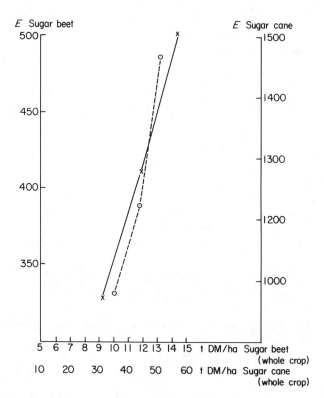

FIG. 12.3   Effect of yield on efficiency of support energy (SE) use in the production of dry matter (DM). N.B. Because of the limited data available the SE used has not been varied for the different yields. ⊙ sugar cane, Hawaii; × Sugar beet, UK.

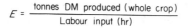

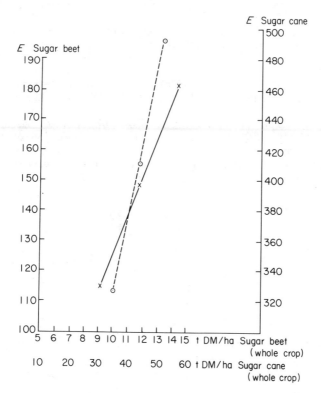

FIG. 12.4 Effect of yield on efficiency of labour use in the production of dry matter (DM). N.B. Because of the limited data available the labour input has not been varied for the different yields.⊙ Sugar cane, Hawaii; × Sugar beet, UK.

References

Austin, R. B., Kingston, G., Longden, P. C. and Donovan, P. A. (1978). Gross energy yields and the support energy requirements for the production of sugar from beet and cane; a study of four production areas. *J. Agric. Sci., Camb.* **91**, 667–75.

Chapman, H. L., Kidder, R. W., Kirk, W. G., Haines, C. E., Casselman, T. W. and le Grand, F. (1964). Proc. Soil and Crop Science Soc. of Florida, **24**, 486–497.

Coombs, J. (1975). Total utilization of the sugar cane crop. Paper presented to the Joint Meeting — British Photobiology Soc. and Int. Solar Energy Soc. (UK Section) Solar Energy: Biological Conversion Systems. 25–27. Imperial College, London.

Cooper, J. P. (1970). Potential production and energy conversion in temperate and tropical grasses. *Herb. Abstr.* **40**, **(1)**, 1–15.

Crane, E. (1975). The flowers honey comes from. *In* "Honey", (Ed. E. Crane), Heinemann, London.

Heichel, G. H. (1973). Comparative efficiency of energy use in crop production. Bull. 739. Connecticut Agricultural Experiment Station, New Haven.

Hudson, J. C. (1975). Sugarcane: its energy relationships with fossil fuel. *Span* **18(1)**, 12–14.

International Sugar Organization (1976). "Sugar Yearbook", London.

M.A.F.F. (1975). Tables of Feed Composition and Energy Allowances for Ruminants.

Nix, J. (1976). "Farm Management Pocketbook", 7th edn, Wye College, Kent.

Paturau, M. J. (1969). "By-products of the Cane Sugar Industry", Elsevier, Amsterdam.

Scott, R. K. (1977). The growth of potatoes and sugar beet. *Agric. Prog.* **47**, 139–153.

13

OTHER CROP PRODUCTS

The range of crop products is enormous and, even though the most important have been dealt with in the preceding chapters, a vast number have not
even been mentioned. In many cases, it is difficult to imagine any purposes
for which comparative efficiencies would be calculated, although, within a
product, there are often alternative processes, some of which are more
costly of one resource or another. In some cases, large inputs of labour are
involved: in others, energy-expensive distillation and extraction processes
are concerned with the production of quite minute quantities. It is instructive to outline the main categories of these products.

Spices

The major spices, such as peppers, cinnamon, cloves, vanilla and mints, may
be thought of as non-essential nutritionally but it would be unwise to dismiss
them as wholly unimportant in making food palatable, digestible and attractive: indeed, it could be argued that they have an important role in the
production of *food* from agricultural raw materials. However, it is hard to
see how efficiency calculations can greatly aid in the understanding of such
processes. Yields from spice crops are usually low. For example, the yield of
peppermint oil (mainly menthol), used for antiseptics, lotions, dentifrices,
medicinals, and culinary flavours, is of the order of 34 kg of distilled oil per
hectare (Janick *et al.*, 1974).

Spices are mostly of ancient origin, reflecting their role in food preservation as well as flavouring.

Perfumes

These probably represent the smallest crop yields that can be imagined as
ever having significance. Most perfumes are based on aromatic essential oils

(benzene or terpene derivatives) obtained from flowers: essences used in soaps and cosmetics are usually derived from vegetative parts.

Oils are extracted by enfleurage (absorption into cold fat), distillation, solvent extraction or "expression" (squeezing). The addition of fatty oils and resins (including ambergris, a fatty substance from the sperm whale) is used to extend and preserve these expensive oils. Rose oil (from *Rosa* spp.) is derived by distillation, enfleurage or solvent extraction and about 1 kg of flowers produces <0.5g of oil: the flowers are generally harvested by hand.

Other essential oils

Other essential oils with important industrial applications, include camphor, from the camphor tree (*Cinnamomum camphora*), used in medicines and in the making of celluloid, and turpentine (from *Pinus* spp.), with a whole range of uses. It has been estimated (Janick *et al.*, 1974) that up to 20 l of crude turpentine may be obtained for each 1017 kg of air-dried wood pulp.

Medicinals

Medicinals also include castor oil, belladonna, digitalis, cascara, cocaine, opium, olive oil, clove oil, betel, nicotine and gum arabic, all well known for hundreds of years, but the discovery of new principles continues, with antibiotics such as penicillin and growth-regulators such as gibberellin being relatively recent examples.

The problems of assessing the efficiency with which they are produced are real ones, as exemplified by the commercial needs of plant to produce penicillin. It is idle to suppose that because a product has remarkable life-saving properties it does not have to be produced efficiently: whoever produces it also has to make a living. Equally, it is hard to estimate the real value of insecticides like pyrethrum and medicinals such as quinine. (The significance of a substance like strychnine, mainly from the seed of *Strychnos nux-vomica*, defies the imagination.)

Masticatories and fumitories

These are another group with complicated social significance. Many leaves have been chewed for pleasure (e.g. Coca) and the betel nut, the seed of the betel palm (*Areca catechu*), is chewed by vast numbers of people in Southeast Asia.

Quantitatively the most important of the fumatories, tobacco (*Nicotiana tabacum*) is also chewed, as is gum (guttagum, from *Conva* spp.), but most tobacco is smoked. Tobacco cultivation is labour intensive and, as is well known, sustains an enormous industry.

Tannins and dyes are nowadays of much less importance than in the past, but the significance of tannins was and is very high in relation to leather production.

The main species used for tannin production are given in Table 13.1, which also illustrates the very high contents of tannin achieved by many of them.

TABLE 13.1   The main species used for tannin production

Plant		Part used	Tannin content[a] (%)
Mangrove	*Rhizophora* spp.	Bark	34
Cassia	*Cassia auriculata*	Bark	
Divi-divi	*Caesalpinia coriaria*	Pods	50
Gambir	*Uncaria gambir*	Leaves	40
Quebracho	*Schinopsis lorentzii*	Heart wood	20–30
Terminalia	*Terminalia chebula*	Pods and bark	
Wattle	*Acacia* spp.	Bark	11–50
Canaigre	*Rumex hymenosepalus*	Taproot	35
Sumac	*Rhus* spp.	Dry leaves	35
Hemlock	*Tsuga canadensis*	Bark	50
Oak	*Quercus aegilops*	Acorn cups	45

[a]These references do not indicate whether these percentages refer to wet or dry material

Sources: Janick *et al.* (1974); McIlroy (1963)

Resins, gums and other exudates

These too have been largely replaced by synthetics. Amber (from the extinct *Pinus succinifer*) is still valued as an ornament, agar from seaweeds is still commonly used as a culture medium for micro-organisms, pectins are used in jellies, and gum arabic (from *Acacia* spp.) features amongst the medicinals.

Latex products

Latex derivatives are by far the most important of the exudates, rubber being the best known. The rubber tree (*Hevea brasiliensis*) still produces about 3.5 million metric tonnes annually, although synthetics have been cheaper. Higher energy and labour costs of synthetics has resulted in some reversal of this trend: the relative energy consumption figures are shown in Table 13.2.

TABLE 13.2 The support energy inputs required to produce natural and synthetic rubber (after UNIDO/UNEP, 1975)

	Energy input required per ton (GJ)
Synthetic latex[a]	154
Natural rubber[b]	13.2

[a]Energy content of raw feedstock plus energy used in processing
[b]Energy used to grow trees (including fertilizers), to harvest and process the rubber

Beverages

Amongst the crop products dealt with in this chapter, the beverages have the greatest nutritional significance. They lie part way between the food crops and those consumed for "non-nutritional" pleasure. The major beverages can be grouped as (a) alcoholic and (b) non-alcoholic.

(a) *Beers, wines and spirits*

Beers are produced by the fermentation of carbohydrates by yeast (*Saccharomyces* spp.) and are usually flavoured by the addition of hops (*Humulus lupulus*). Local beers may be based on potatoes (chicha), rice (sake), agave (pulque) or honey (mead), but the malt beers are based on barley.

Although beers are not produced or consumed simply for their energy content, this is not negligible and it is interesting to compare the output of energy in the barley crop and that in the beer made from it (Table 13.3). This takes no account, of course, of the additional support energy involved in making beer.

TABLE 13.3   The energy content of barley and the beer made from it

		Energy content (MJ)
Inputs		
Barley from 1 ha (3 767 kg)		59 274
Other ingredients, e.g. corn syrup and caramel		12 196
	Total	71 470
Outputs		
177 barrels of beer (28 967 l)		52 141
Brewers' grains (860 kg DM)		16 856
	Total	68 997

Source: Harrison (1978)

Wines also can be made from a variety of crop products, but mostly they result from the fermentation (with the yeast *Saccharomyces ellipsoideus*) of crushed grapes, with or without the skins. They are more alcoholic than beers, which contain about 90% water, with 10–14% of alcohol and 70% of water.

Spirits are produced by distillation of the equivalent of beer or wine, and condensation and fractionation of the resulting alcohol: the final product contains about 50% of alcohol.

Alcohol has many industrial uses but for these purposes is produced by the distillation of wood.

The output of alcohol in the form of beers, wines and spirits, per unit of land, crop and support energy, is illustrated in Table 13.4

(b) *Non-alcoholic beverages*

Apart from increasing quantities of "soft" fruit drinks, the major non-alcohol beverages are coffee, tea, cocoa and maté.

They are derived from berries or leaves and are mainly harvested by hand: the labour cost per unit of product is therefore high (Table 13.5). Most of these beverages contain caffeine and have stimulatory properties.

TABLE 13.4   The output of alcohol in the form of beer, wine or spirit

	Output of alcohol per annum (litres)		
	per hectare of land	per tonne of crop	per megajoule of support energy
Beer	955	209	0.10
Wine			
Europe	750–1 050	150–210[a]	
California	1 500–2 100	300–420[a]	
Whisky	1 426–1 491	386–403	0.002

[a]These figures are for the wet weight of the crop which has a dry matter content of approx 35% compared with approximately 86% for the grains used in beer and whisky production. If the calculations for wine are made on a tonne of crop dry matter, the yields of alcohol are of the order of 440–620 litres per tonne of crop.

TABLE 13.5   Labour requirement for non-alcoholic beverages

		Total labour (hr ha^{-1})	Yield (kg ha^{-1})		Labour (hr kg^{-1})
Coffee	Brazil estates	576	446	(green beans)	1.3
	Columbia smallholdings	799	523	(green beans)	1.5
	El Salvador estates	1 566	660	(green beans)	2.4
	Kenya estates	2 615	840	(green beans)	3.1
	Ivory Coast	840	1 121	(dried beans)	0.7
Tea	Sri Lanka	4 495			
	Kenya	4 729			
Cocoa	Ivory Coast	930	1 000	(dried beans)	0.9

Source: Ruthenberg (1976)

Maté, or Paraguayan tea, is made from the leaves of the maté tree (*Ilex paraguayiensis*), guarana from the seeds of *Paullinia cupana*, and cola from the seeds of the cola tree (*Cola ritida*). The major beverages of the world are coffee, made from the beans of the coffee tree (*Coffea* spp), tea, from the leaves of the tea plant (a bush, *Thea sinensis*), and cocoa and chocolate, from the beans of the understory tree (*Theobroma cacao*). The relative yields of these products are given in Table 13.6 and an indication of the energy and fertilizer inputs in Tables 13.7 and 13.8.

TABLE 13.6   Yields of established crops of coffee, tea and cocoa

		Yield (kg ha^{-1})	Sources[a]
Coffee	Hawaii	2 242	(1)
	Costa Rica	953	(1)
	Kenya	1 004	(1)
	Kenya	840	(2)
	El Salvador	807	(1)
	El Salvador	660	(2)
	Colombia	504	(1)
	Colombia	523	(2)
	Brazil	404	(1)
	Brazil	446	(2)
	Ivory Coast	1 121	(2)
Tea	Darjeeling	504	(1)
	Sri Lanka (average)	1 121–1 457	(1)
	Sri Lanka (best)	over 2 242	(1)
	Assam	1 233–2 242	(1)
	Sri Lanka 1964–65	910	(2)
	Sri Lanka 1972	800	(2)
	Sri Lanka — intensive production	3 000–4 000	(2)
Cocoa	Trinidad (average)	224	(1)
	Trinidad (selected cuttings)	673–1 121	(1)
	Trinidad (best strains)	2 242	(1)
	West Africa	224–3 363	(1)
	Ivory Coast	1 000	(2)

[a]Sources: (1) Purseglove (1968); (2) Ruthenberg (1976)

N.B. Yields are expressed in different terms in the literature e.g. clean coffee, dried beans, made tea, dry cocoa, green coffee etc. so it is difficult to be sure of comparability

TABLE 13.7 Support energy inputs to provide tea, coffee and chocolate (cocoa) in the United Kingdom (after Chapman, 1975)

	Support energy input (MJ kg^{-1})
Tea	150.8
Coffee	309.4
Chocolate	63.5

TABLE 13.8 Examples of fertilizer inputs for coffee, tea and cocoa crops

	N	Fertilizer input (kg ha^{-1} yr^{-1}) P_2O_5	K_2O
Coffee	135	34	145
Ecuador (shaded)	100	50	100
Ecuador (unshaded)	150–200	75–100	150–200
Tea			
N. India	90–135	7	13
Ceylon (seedling)	90–225	22.5	45–100
Ceylon (clonal)	270–405	35	100–135
S. India	80–200	33.5	40–67
Cocoa			
Nigeria (young trees)	65–100	50	
Nigeria (mature trees)	65–135	50–65	

Source: Geus (1973)

N.B. As with other crops fertilizer requirements depend on the soil conditions, productivity of the crop, and, for these crops, whether or not they are shaded.

Within each product, it is possible and often useful to calculate the efficiency with which such resources are used but there is little to be gained from comparisons between products, unless one was concerned, for example, with the output of caffeine.

References

Chapman, P. (1975). "Fuel's Paradise: Energy Options for Britain", Penguin Books, Harmondsworth, Middlesex, UK.

Geus, J. G. de (1973). "Fertilizer Guide for the Tropics and Subtropics", 2nd edn, Centre d'Etude de l'Azote, Zurich.

Harrison, L. E. (1978). Personal communication.

Janick, J., Schery, R. W., Woods, F. W. and Ruttan, V. W. (1974). "Plant Science", 2nd edn, W. H. Freeman, San Francisco.

McIlroy, R. J. (1963). "An Introduction to Tropical Cash Crops". Ibadan University Press, Nigeria.

Purseglove, J. W. (1968). "Tropical Crops. Dicotyledons 2", Longmans, London.

Ruthenberg, H. (1976). "Farming Systems in the Tropics", 2nd edn, Clarendon Press, Oxford.

U.N.I.D.O/U.N.E.P. (1975). Report of the expert group meeting on the study of synthetic versus natural products. Pilot project on the rubber industry and its impact on the environment. ID/WG.188/3.

14

FIBRE PRODUCTION

Vegetable fibres may be classified in several ways: Table 14.1 groups them according to the part of the plant from which they are obtained.

TABLE 14.1   Classification of vegetable fibres according to the part of the plant from which they are obtained (after Kirby, 1963)

Part of plant	Commercially important fibres
Seed or fruit	Cotton, kapok, Akund floss and other flosses or silk-cottons
Inner bast tissue or bark of stem	Flax, hemp, jute, ramie
Leaves	Abaca or Manila hemp, sisal, henequen, Mauritius hemp, *Phormium tenax*
Woody fibres of trees	Trees used for paper making
Sheathing leaf-stalks of palm trees	Piassava and similar brush-making fibres
Husk or mesocarp	Coir
Roots	Mexican whisk

They can also be classified according to their uses, but these overlap somewhat. Nevertheless, since this represents the possibility of alternatives for a given product, they are of some importance.

Broadly speaking, cotton, ramie, jute and flax are mainly textile fibres, used for making fabrics, clothing and bags, whilst sisal, abaca, henequen, hemp and flax are mainly cordage fibres, used chiefly for ropes and twines.

In both of these groups fibre can be spun, but this is also true of coir, which is generally classed amongst the brush and mat fibres. Stuffing and upholstery materials form another important group and there are also miscellaneous fibres, used for tying (e.g. raffia) and basket making. Finally, there are the materials used for paper making: these are mainly wood, especially the soft-woods, but include some grasses, cotton, flax and abaca.

One of the features distinguishing fibres from plant food products is that strength matters. This is the most important single quality for most fibres, tensile strength being measured as the breaking load per unit area of cross-section, expressed in terms of kg mm^{-2} (Table 14.2). (A more specific term often used to describe tensile strength measured in this way is "tenacity".)

TABLE 14.3   Cellulose content of vegetable fibres

	Cellulose (%)	Hemicellulose	Sources[a]
Flax (retted)	64	17	(1)
Hemp	67	16	(1)
	70	—	(2)
Straw Barley	43.2	30.3	
Oat	44.6	28.2	(4)
Wheat	40.3	29.2	
Rice	36.2	14.5	
Kapok	64	—	(2)
Cotton lint	94	—	(2)
Sisal[b]	78	10	(3)
Abaca	63.2	19.6	(1)

[a]Sources: (1) Kirby (1963); (2) Purseglove (1968); (3) Lock (1969); (4) Owen (1976)
[b]62% "true" cellulose, 16% pentosan and 10% other carbohydrates, hemicelluloses, pectins, etc.

Other properties are also important, such as elasticity, stiffness, resilience, bending qualities, resistance to fatigue, and fineness (important for spinning). Most plant fibres contain a high proportion of cellulose (Table 14.3) but the useful fibres have to be extracted in many cases by processes of soaking (retting), washing and cleaning. All these processes greatly add to the labour requirements of fibre production (Table 14.4). These are very high for the stem fibres.

TABLE 14.2 Strength of some plant fibres (after Stout and Jenkins, 1955)

	Area of cross-section[a] ($cm^{-2} \times 10^{-5}$)	Tensile strength[b] ($kg\,mm^{-2}$)	Extension at break (%)	Tensile strength % extension (elastic modulus)
Jute *Corchorus* sp.	1.2	105	2.7	38.9
Sisal *Agave sisalana*	23.2	78	5.0	15.6
Flax *Linum usitatissimum*	1.2	134	4.1	32.7
	0.3	183	3.2	57.2
Hemp *Cannabis sativa*	0.6	126	4.2	30.0
Ramie *Boehmeria nivea*	0.3	129	3.9	33.1
	0.8	112	4.2	26.7
Kenaf *Hibiscus cannabinus*	2.1	87	3.5	24.8
Roselle *Hibiscus sabdariffa*	2.8	91	3.5	26.0

[a] of fibre substance
[b] Corrected to standard time to break of 10 seconds

TABLE 14.4   An example of the
breakdown of labour requirements
for sisal production (Tanganyika,
1956, after Kirby, 1963)

Process	Approx % of total labour requirement
Development	17
Cultivation	29
Cutting	25
Factory	27

Yields vary with soil conditions and, in some cases, depend greatly on rainfall, but may be extremely high, especially where three or even four crops (e.g. of ramie) may be taken per year (Tables 14.5 and 14.6). Only a small proportion of the green plant is actually harvested as fibre (Table 14.6), depending upon the efficiency of the extraction process.

TABLE 14.5   Yields of fibre crops

Crop	Yield (kg ha^{-1})	Sources[a]
Cotton lint	112–2 242	(1)
(California-irrigated)	up to 3 360	(1)
Flax fibre	500–1 000	(2,4)
Sisal fibre	1 120–2 800	(1,2)
Abaca fibre	1 255–5 022	(2)
Henequen fibre	2 033	(1)
Hemp fibre	560–2 500	(2)
Jute fibre	852–5 605	(2)
Jute fibre	1 200–1 800	(3)
Kenaf fibre		
(usual)	1 121–3 363	(2)
(exceptional)	6 726	(2)
Ramie fibre	627–3 139	(2)
Kapok		
(10-year old trees)	800–1 600	(3)
(mature trees)	2 000–4 000	(3)

[a]Sources: (1) Purseglove (1968); (2) Kirby (1963); (3) Geus (1973); (4) Dempsey (1975)

TABLE 14.6   Proportion of green crop harvested as fibre

Crop	Green crop (kg ha^{-1})	Fibre as % of green crop	Sources[a]
Flax	6 000	16–24	(1)
Sisal	75 000	3.5–5	(2)
Abaca			
Musa textilis	251 000	2–3	(2)
Henequen			
Agave fourcroydes	50 800	4	(3)
Hemp	10 100–12 300	24–25	(2)
Jute	23 500–34 700	4–8	(2)
Kenaf	168 000	4	(2)
Ramie	45 000–55 000	3.5	(1)
	50 200	2.5	(3)

[a]Sources: (1) Dempsey (1975); (2) Kirby (1963); (3) Purseglove (1968, 1972)

Two results of all this are that (a) a great deal of fertilizer is required (Table 14.7) to sustain the enormous yields of green matter and (b) a great deal of organic matter remains after the fibre has been extracted. This residue is frequently used as a fertilizer and considerably reduces the quantity of other fertilizer that has to be added.

TABLE 14.7   Fertilizer requirements of fibre crops other than cotton

Crop	Fertilizer (kg ha^{-1}) N	P$_2$O$_5$	K$_2$O	Sources[a]
Flax	20–50	30–120	up to 140	(1)
Jute	80–120	20–40	8–180	(1)
Sisal	30–100	62–124	250–500[b]	(2)
Hemp	40–150	30–100	40–120	(1)
Ramie	up to 500	20–180	60–200	(1)
Kenaf	up to 150	up to 134	up to 200	(1)
Roselle	22–80	up to 40	up to 120	(1)

[a]Sources (1) Dempsey (1975); (2) Geus (1973)
[b]kg KCl per hectare per cycle

However, all the inputs that are related to the large yields of green matter, appear rather high in relation to the actual fibre production. Of course, in the case of coir or coconut fibre, the main product is the fruit and the fibre is a by-product: it therefore only has to bear the additional cost of the resources used to harvest and process it.

Not all fibre production is at a high level per unit of land. The kapok tree (*Ceiba pentandra*), for example, begins to bear in its third or forth year but, at this stage, its yield is about 100 pods, giving about 130 kg of cleaned floss per hectare. From mature trees a yield of about 600 kg ha^{-1} is considered excellent. Harvesting is laborious and involves climbing the trees.

The most important of the floss-from-seed plants is cotton (*Gossypium* spp.). The crop is grown in many countries of the world but quantitatively the most important are the USA, USSR, China and India.

The cotton plant is a perennial and requires a hot, relatively dry climate. The cotton fibres are attached to the seed, so the yield per plant is related to successful flowering and seed setting.

The proportion of the whole above-ground plant actually represented in cotton fibres is small by weight (Table 14.8) but the yields per hectare can be substantial (Table 14.9). The crop is labour-intensive (Table 14.10), depending upon the degree of mechanization, but an efficient user of support energy per unit weight produced (Table 14.11) (compared with animal or synthetic fibres).

TABLE 14.8   Cotton fibre weight as proportion of total above-ground plant[a]

Part of plant	Dry weight (kg ha^{-1})
Stems	1 228
Leaves	1 074
Burs	754
Seed	1 221
Lint	560
Total above ground	4 837
Lint as % of total = 11.6%	

[a]Based on a yield of 560 kg of lint per hectare

Source: Christidis and Harrison (1955)

TABLE 14.9 Average yield of cotton fibre per hectare in 1971 for some countries

Country	Yield (kg ha^{-1})
Greece	900
USSR	840
El Salvador	860
Egypt	780
Israel	1 190
Burma	110
Kenya	70
USA	490

Obtainable yields

California-irrigated
  up to 3 360
  average 1 120–1 680

Sources: F.A.O. (1972); Berger (1969); Purseglove (1968)

TABLE 14.10 Labour requirements for cotton production

	hours per hectare	hours per bale[a]	Yield (kg ha^{-1})	Sources[b]
USA 1970–72	59	25	512	(1)
Estimated requirement without extensive mechanization	370			(2)
USA 1968		49		(3)

[a]Wt. of bale = 480 lbs (218 kg)
[b]Sources: (1) U.S.D.A. (1973); (2) Janick et al. (1974); (3) Berger (1969)

TABLE 14.11   Support energy used in the production of various fibres (after Reid and White, 1977)

Fibre	Raw materials	Support energy used[a] (MJ kg^{-1})
Cotton	Soil nutrients, CO_2, solar energy	29.5[b]
Wool	Rangeland cellulose and nitrogen	76.6[c]
Synthetic cellulosics	Wood cellulose and chemicals	188.3
Synthetic non-cellulosics, e.g. polyester, nylon	Petrochemicals	140.3

[a]Includes energy cost of labour, machinery depreciation and maintenance, electricity and transportation fuels.
[b]Energy inputs to cotton production are divided between cotton seed oil, cotton seed meal and cotton fibres.
[c]Includes an appropriate proportion of the energy inputs to sheep production plus the energy inputs to shearing, transportation and scouring of wool.

Fertilizer requirements (Table 14.12) are not especially high but irrigation may be needed in some circumstances.

TABLE 14.12   Fertilizer requirements for cotton (after Geus, 1973)

	Fertilizer (kg ha^{-1})		
	N	P$_2$O$_5$	K$_2$O
USA (1970 usage)	28–150	30–95	15–100
USA irrigated	45–330		
Brazil	30–45	30–90	30–70
India			
Madhya Pradesh local varieties (heavy soils)			
irrigated	100–120	60	60
rain-fed	60–80	40	20
Hybrid-4			
heavy soils	160–200	80	60
light soils	40–60	30	20

As with most such fibres, the product tends to be almost unique. Other fibres, or even quite other substances, can be substituted for any of the functions actually performed, but if cotton is wanted, then no other product will do quite as well. This is not a unique proposition for fibres, of course, but is more commonly the case for these than for the major food crops.

The main interest in efficiency, therefore, lies in the alternative ways of producing, harvesting and processing the crop, and the ways in which yields can be increased and losses decreased.

Paper-making Fibres

Paper consists of near-pure cellulose fibres and the manufacturing processes are both capital- and energy-intensive (Table 14.13). This also applies to the pulping processes that are involved in turning wood into paper — and most paper comes from wood, chiefly softwoods, such as spruce (*Picea excelsa*) and pine (*Pinus sylvestris*).

TABLE 14.13　Capital and energy inputs to fibre production in USA (1972)

Fibre	Production (short tons × 10⁶)	Capital input ($ per ton)	Energy purchased (J × 10⁹ ton⁻¹)
Paper[a]	57.5	15.5	21.5
Cotton[b]	2.7		24.2

[a]Jahn and Preston (1976)
[b]US Cotton Handbook (1978)

The yield of pulp (largely cellulose) from wood is about 45–55%.

The yield of wood can be high per unit of land (Table 14.14), per unit of labour (Table 14.15), depending upon the degree of mechanization, and per unit of support energy, but processing and manufacturing may result in costly products per unit of support energy (Tables 14.16 and 14.17).

Wood as a Material

Timber has many advantages as a crop. It can be very productive per unit of land (Table 14.18), the product is relatively dry and easy to store (before or

TABLE 14.14   Yield of wood for pulp and papermaking

Species	Conditions	Mean annual increment (m^3 ha^{-1})
Conifers	North temperate zones and some Mediterranean countries	2–5
Conifers	Extremely favourable	10–20
Conifers	Tropics and subtropics	15–20
Eucalypts and Poplars		20–30

Source: F.A.O. (1973)

TABLE 14.15   Examples of yields of wood per unit of labour

	Yield of Wood (Tonnes per man per year)		
	No mechanization	Partial mechanization	Full mechanization
African forests (French territories overseas, 1949)	15	60	100

Source: O.E.E.C. (1950)

TABLE 14.16   Yield of wood products per unit of support energy

Product	Country	Source	Yield oven-dry wood (kg) per MJ support energy
Softwood lumbering	USA	Jahn and Preston (1976)	0.15[a]
Oak flooring	USA	Jahn and Preston (1976)	0.13[a]
Insulation board	USA	Jahn and Preston (1976)	0.08[a]
Eucalypt chips	Australia	Saddler (1975)	0.13[b]

[a] Includes energy inputs to extraction and manufacturing but not growing trees
[b] Includes energy inputs to growing, harvesting and transporting

TABLE 14.17   Support energy inputs[a] to paper production (after Reding and Shepherd, 1975)

	Energy input (KJ per kg paper)
Debarking/chipping logs	Bark used as fuel[b]
Digestion of wood chips	2 900
Evaporation of water from digestion	2 120
Calcination	1 970
Drying	4 400

Yield of paper per MJ support energy is 0.09 kg

[a]The figures are for modern, well-operated plants and do not include the bleaching and coating operations.
[b]Process wastes including bark are used to generate the electricity used in debarking/chipping logs.

TABLE 14.18   Yields of wood for timber (after Logan, 1967)

Species	Yield[a] ($m^3\ ha^{-1}\ yr^{-1}$)
Eucalypts and poplars	20–30
Eucalypts and poplars (good sites and irrigation)	up to 40
Gmelina and Maesopsis eminii	10–15
Teak	6–10

[a]Yields may vary enormously according to conditions e.g. under best conditions, increments, both of height and volume, may be three or more times those on the poorest site.

after harvesting) until required. As a living store, it will continue to grow and accumulate until the point of harvest, allowing harvest time to be selected on grounds other than that the plant or product will deteriorate.

As a material, wood can be very long-lasting and its energy content is not necessarily lost by material use.

References

Berger, J. (1969). "The World's Major Fibre Crops", Centre d'Etude de l'Azote, Zurich.

Christidis, B. G. and Harrison, G. J. (1955). "Cotton Growing Problems", McGraw-Hill, New York.

Dempsey, J. M. (1975). "Fiber Crops", A University of Florida Book. The University Presses of Florida, Gainsville.

F.A.O. (1972). Production Yearbook 1971. Vol. 25, F.A.O., Rome.

F.A.O. (1973). Guide for planning pulp and paper enterprises. F.A.O. Forestry and Forest Products Studies No. 18. F.A.O., Rome.

Geus, J. G. de (1973). "Fertilizer Guide for the Tropics and Subtropics", 2nd edn, Centre d'Etude de l'Azote, Zurich.

Jahn, E. C. and Preston, S. B. (1976). Timber: more effective utilization. *Science N.Y.* **191** (4228), 757–761.

Janick, J., Schery, R. W., Woods, F. W. and Ruttan, V. W. (1974). "Plant Science", 2nd edn, Freeman, San Francisco.

Kirby, R. H. (1963). "Vegetable Fibres", Interscience, N.Y.

Lock, G. W. (1969). "Sisal", 2nd edn, Longmans, London.

Logan, W. E. M. (1967). F.A.O. World Symposium on Man-Made Forests and their Industrial Importance. 1. Policy. *Unasylva* **21(3–4)**, Nos. 86–87, 8–22.

O.E.E.C. (1950). Possibilities of increasing the use of tropical timber. Report of a group of experts.

Owen, E. (1976). Farm wastes: straw and other fibrous material. *In* "Food Production and Consumption", (eds A. N. Duckham, J. G. W. Jones and E. H. Roberts), North Holland. Publishing Co., Amsterdam.

Purseglove, J. W. (1968). "Tropical Crops. Dicotyledons. Vols 1 and 2", Longmans, London.

Purseglove, J. W. (1972). "Tropical Crops. Monocotyledons Vols 1 and 2", Longmans, London.

Reding, J. T. and Shepherd, B. P. (1975). Energy consumption: paper, stone/clay/glass/concrete, and food industries. Dow Chemical Co. PB-241 926. Prepared for US Environmental Protection Agency.

Reid, J. T. and White, O. D. (1977). Energy utilization and conservation in animal production. *In* "Energy Conservation in Agriculture", Special Publication No. 5. Council for Agricultural Science and Technology, Iowa, USA.

Saddler, H. D. W. (1975). Organic wastes and energy crops as potential sources of fuel in Australia. Energy Research Centre, School of Biological Sciences, University of Sydney.

Stout, H. P. and Jenkins, J. A. (1955). Comparative strengths of some bast and leaf fibres. *Ann. Sci. Text Belges* No. 4, 231–251.

US Cotton Handbook (1978). Cotton Council International and the National Cotton Council of America.

U.S.D.A. (1973). "Agricultural Statistics 1973", U.S.D.A. Washington D.C.

15

FUEL CROPS

The most important fuel crop by far is wood. It has been so in the past, for most of the world, and, as non-renewable resources become exhausted or too costly to obtain, it could increase in importance in the future (National Academy of Sciences, 1980).

In 1970 world energy consumption in the form of wood and charcoal was 487 million tonnes coal equiva'ent, compared with 2419 for coal, 2850 for oil and 1418 for natural gas (a total of 7435 million tonnes coal equivalent). It has been estimated (F.A.O., 1971) that over half of the world's annual wood production is utilized as fuel, 90% of it being in the developing world.

However, only about 13% of the world's forest increment is currently harvested and only 6% of it for use as fuel, and Earl (1975) has estimated that the forest resource is "theoretically large enough to cater for human energy needs without affecting the supply of wood for industrial purposes".

The economics of such provision depend greatly on whether the demand is close to the source of supply and how far wood has to be transported. The bulkiness of wood is one of its main disadvantages, for both storage and transport: it is also labour intensive. However, it stores easily and safely even for long periods, and can be used safely by unskilled labour.

The energy or fuel value of wood depends upon its moisture content (Table 15.1) but it is a lower than that of the fossil fuels (Table 15.2).

A combustion efficiency (CE) is also often calculated, as follows:

$$CE = \frac{NHV - [W_g \times \text{sp.ht} \times (t_2 - t_1)] + L_r}{FV} \times 100$$

where CE is the combustion efficiency (%), NHV the net heating value of incoming wood (Btu $1b^{-1}$), W_g the weight of stack gas (1b), sp. ht the specific heat, t_1 the temperature of incoming air (°F), t_2 the temperature of

185

TABLE 15.1 Energy value of wood at different moisture contents (after Tillman, 1978)

Moisture content (as-received basis) (%)	Fuel value softwood (MJ kg^{-1})
0.0	20.5
16.7	16.6
28.6	13.8
37.5	11.7
44.4	10.1
50.0	8.8
54.5	7.8
58.3	6.9
61.5	6.1
64.3	5.5
66.7	4.9

TABLE 15.2 Energy value of fossil fuels and peat (after Earl, 1975)

Fuel	Energy value (MJ kg^{-1})
Paraffin	43.5
Fuel oil	29.0
Coal (bituminous)	28.9
Peat (air dry)	16.7

stack gas (°F) and L_r the radiation loss (normally ~ 4%) (Tillman, 1978). (1 btu = 1.055 kJ; 1 lb = 0.455 kg). Fuel values are calculated either on a gross basis, which includes the heat derived from the condensation of any water vapour produced by combustion, or on a net basis, which excludes heat obtained by condensation. This is an important difference for many purposes. The heat of condensation cannot always be recovered and in many fuel-burning situations wet wood simply does not provide as much heat, since a proportion of that produced is used to evaporate the water which condenses elsewhere. This proportion can be substantial and, for bark, for example, unless the moisture content is less than 88% the fuel value is actually negative. By itself, therefore, it will not burn at all.

Fortunately, simple storage allows wood, and bark, to dry out substantially, without any additional energy expenditure, due to the capacity of the atmosphere to absorb moisture. In fact, of course, the necessary additional energy comes from the sun; this represents quite an important way in which more use can be made of solar radiation. Support energy is used in forest production (Table 15.3) but most of the inputs occur after harvesting and during processing.

TABLE 15.3   Support energy usage in fuelwood production from a 10 000 ha forest where 360 ha are clear-felled each year (after Cousins, 1975)

	Support energy usage (MJ per year)
Sowing and planting trees	negligible
Machinery	
manufacture	4.15×10^6
fuel used	1.88×10^7
Transport in logging trucks	8.46×10^6
Chipping	
machinery manufacture	1.84×10^5
fuel used	1.84×10^6
Fire protection	
chemical production	2.90×10^6
application	1.90×10^6
Total input	36.52×10^6
Yield of wood	1.84×10^9 MJ per year
Energy input per MJ wood produced	0.02 MJ

One widespread form of processing for fuel is conversion to charcoal. This is a controlled carbonization involving combustion in kilns or furnaces, during which about 55% of the energy in the wood is lost. The product has several advantages, however, being smokeless as a fuel and useful for a number of industrial processes. Its fuel value is similar to that of coal and it acts as a strong reducing agent when heated.

Charcoal may also be produced by wood distillation, in retorts from which the condensed gases may be fractionated to produce methyl alcohol, acetic acid and pitch.

The yield of charcoal by weight is usually between 20% and 30% of the dry weight of wood, and by volume about 50%. Moisture content varies from 1% to 16% and fixed carbon content from 80% to 90% (Earl, 1975); the fuel value is about 29.7 kJ g^{-1}.

This is another example of conversion of one raw material (or product) into another with different properties, but it is also an example of increased concentration, achieved at some cost, allowing a reduction in the costs of further handling and transport. The efficiency of the total process must depend, of course, on such factors as method of transport and distance to be travelled.

Another naturally produced and long-established fuel crop is peat. Peat results from the accumulation of undecomposed sphagnum moss (*Sphagnum* spp.) in the anaerobic (and usually acidic) conditions of open bogs and marshes. Extensive areas of these bogs occur in countries such as Eire (6% of its land surface) and Finland and both are investigating the possibilities of making greater use of peat as an industrial fuel. Air-dry peat has a fuel value similar to that of wood (see Table 15.2) but it is very laborious to cut (although machinery now exists to do this) and it takes a considerable time to produce a harvestable crop (see Table 15.4 for annual growth rates). However, other plants can grow in boggy areas and some of them might prove more productive (Lalor, 1975).

TABLE 15.4 Estimated annual growth of peat (after Clymo, 1978)

	Growth rate $(g\ m^{-2}\ yr^{-1})$
Glenamoy (Ireland)	32
Moor House (UK)	180

A great deal of useful fuel may be derived from crop residues and by-products, notaby cereal straw, which has the advantage of already being harvested (even if it is immediately discarded again by combine harvesters) and of being very dry (Table 15.5). The amounts of cereal straw available are very large: that produced in the UK has been calculated to contain sufficient energy for approximately 40% of the needs of UK farms (Leach, 1975) and it has been estimated (Lalor, 1975) that Australia could generate more than 10% of her electricity by burning her waste wheat straw alone.

Increasing attention is now being paid, however, to the possibilities of deliberately growing (and breeding varieties of) crops for the main purpose of energy production. These "energy crops" (Lalor, 1975) might include

TABLE 15.5   Energy and moisture contents of crop residues
(after M.A.F.F., 1975)

	Moisture content (%)	Gross energy content (MJ per kg DM)
Cereal straws		
Barley	14	17.8
Wheat	14	17.7
Rye	14	18.1
Potato haulm	77	17.3
Sugar beet tops	84	15.4

species relatively new to agriculture, such as bamboo, *Spartina townsendii*, and freshwater plants such as water hyacinths (*Eichhornia crassipes*) and algae, and marine weeds and algae. The giant kelp (*Macrocystis pyrifera*) has been suggested (Szetela *et al.*, 1973), since it is claimed to be one of the world's fastest-growing plants.

Levitt (1975) has pointed out that the by-product parts of crop plants could be converted into charcoal with very little additional energy cost. Corn (*Zea mays*), for example, is grown for the grain, for each 2.8 kcal (11.7 kJ) of which an additional 1 kcal (4.184 kJ) of fuel has already been used (Pimentel *et al.*, 1973), but the grain accounts for only 32% of the total carbon in the plant and the roots contain only about 7% (Miller, 1931). There must therefore be about twice as much carbon in the stems, leaves and cobs as is harvested in the grain. Some three-quarters of this carbon is recoverable as charcoal, having used part of the plant energy in the charring process. Thus for every ton (1.016 tonnes) of grain harvested, about 0.66 tons (671 kg) of charcoal should be available from the rest of the plant. Levitt (1975) calculates that 0.24 tons (243 kg) of charcoal would supply the fuel needs (or its equivalent) of the corn crop and 0.42 tons (427 kg) of excess charcoal would remain for each ton of grain.

Other crops could similarly provide a surplus of fuel, over and above that required to produce them. Downing (1975) has calculated that for the croplands of Ontario, Saskatchewan and for Canada as a whole, the ratios of total energy produced (crop + residue + waste) to total energy used were 7.2:1, 10.9:1 and 8.1:1, respectively.

Indeed, a seven-acre (3.2 ha) experimental underwater farm of kelp already exists off San Clemente Island. It is based on an anchored raft, below the surface of the water, secured to the bottom (at about 300 ft (91.4 m)). Calculations assume a 2% conversion efficiency for converting solar radiation into seaweed, 5% conversion efficiency for the production of human food from the seaweed, and 50% conversion efficiency for the production of other products from the seaweed (Chedd, 1975), and suggest that this type of farming could yield enough food for 1100–1900 persons per square km of ocean area cultivated, in addition to producing enough energy to support more than 115 people at current US per capita consumption levels. (It is estimated that there are some 207–259 million square kilometres of "arable surface water" in the world.) Seaweeds contain little cellulose but can be processed to produce methane, fertilizer, ethanol, lubricants, waxes, plastics and fibres; some of which would be needed for the construction of rafts.

Clearly, a great variety of plants could be used, since they do not have to be edible, or even non-toxic, or digestible by animals: it is only total yield of fixed carbon that matters; the form and the way it is partitioned between plant parts is immaterial.

On land the main product would be cellulose and it is interesting that the US Army Natick Laboratories in Massachusetts have isolated some 13 000 micro-organisms that can live on cellulose (Chedd, 1975). They do this by producing cellulases, and one particular fungus (*Trichoderma viride*) is especially good at breaking down even crystalline cellulose. Mutants have already been selected that produce two to four times as much cellulase as the wild type. This particular organism has been shown to be capable of producing glucose from wastepaper, yielding about 50% of the weight of cellulose as glucose.

One of the major advantages of trees, as with all perennials, is that the product is stored where it has grown and can be harvested as required, continuously throughout the year if necessary. Annual crops, by contrast, have to be harvested within a relatively short season, resulting in large peaks of demand for labour and considerable problems of storage, (a) because of the sheer bulk of energy crops and (b) because such crops are not usually in a storable form.

Amongst the tree species suggested specifically as energy crops, hybrid poplars have shown promise in central Pennsylvania. Planted at 3700 trees per acre (9139 per hectare) and producing six to eight harvests per planting, even on marginal land the output is of the order of 120 million Btu per acre per year (313 GJ ha^{-1} yr^{-1}) at a conversion efficiency of 0.6% (Chedd, 1975).

As Graham (1975) has pointed out, any new energy base has to be judged in relation to (i) the capital investment required, (ii) the need for materials, especially scarce ones, and (iii) the social impact, including effects on the environment. The advantages of agricultural energy crops are that they pose no new environmental hazards and represent a massive renewable resource. Rough estimates are that one ton (1.016 tonnes) of dry organic matter can be converted into nearly two barrels of oil or about 10 000 cu. ft (280 m³) of gas. The economics of such production must be satisfactory, of course, but it is hard to make a relevant calculation, having regard to the rate at which some of the costs and prices change: such calculations as have been made suggest that an energy plantation is likely to be competitive with other fuels for thermal electric plants (Szego and Kemp, 1973).

However, the technology largely exists and the conversion efficiency with which annual solar radiation is used currently is of the order of 0.4% (see Table 15.6, and Szego and Kemp, 1973).

TABLE 15.6 Conversion efficiency of solar radiation (after Szego and Kemp, 1973)

Plant	Age of plant (yr)	Location	Estimated solar[a] energy conversion (%)
Alfalfa			
three cuttings per season	1–	US Midwest	0.29
Reed Canary Grass	1–	US Midwest	0.39
Corn			
mature silage	1–	US Midwest	0.41
stalk and ears	1–	US	0.44–0.69
General agriculture	1–	US	0.28–0.85
Sugar cane	?	Louisiana and Florida	1.2
Slash pine (crown and bole)	20+	South-east States of USA	0.24–0.30
Conifers			
Pseudotsuga toxifolia	18–22	England (lat. 51–52° N)	⎫
Pinus nigra			⎬ 0.37
Picea abies			⎭
Sycamore	5	Georgia	0.64

[a]Based on annual average insolation equal to 1300 Btu ft⁻² day (14.8 MJ m⁻² d⁻¹)

The main process by which organic matter can be converted into heat or high energy content fuels have been listed (Lalor, 1975) as (a) combustion, (b) aerobic fermentation, (c) anaerobic fermentation, (d) pyrolysis, (e) chemical reduction and (f) biophotolysis of water.

Direct combustion

Direct combustion normally requires fairly dry material (<10–15% water content) and may give rise to some pollution. The technical feasibility of direct use of heat for power production is established.

Aerobic fermentation

The production of ethyl alcohol by fermentation of materials containing sugars and starches, in the presence of oxygen, is an established process, and ethyl alcohol can be used as a fuel.

A current example is the production of ethanol, mainly from sugarcane, in Brazil. Since 1977, motor fuel in the Brazilian state of São Paulo has contained 15–20% of ethanol and it is planned to extend this to the whole of the country.

Important in this context is the use of the residual fibre (bagasse), left after the cane has been squeezed to extract the juice, as the source of fuel for the ethanol production process. Some 4.6 tonnes of ethanol can be produced per hectare, from a sugar cane DM yield of about 35 tonnes per hectare.

Anaerobic fermentation

Methane production by anaerobic fermentation of sewage is well known, micro-organisms in the liquefied waste feed on the suspended solids converting them to CO_2 and methane.

The application of this idea to energy crops and agricultural wastes is in the early stages of development.

In so far as wastes are used, the method has the advantage of also solving a disposal problem.

Pyrolysis

Pyrolysis involves heating organic matter, in the absence of oxygen, to very high temperatures (500–900°C), for the production of gases or liquids. Charcoal can be produced from wood in this way and other products include methanol, acetic acid and turpentine.

It is established that pyrolysis can be applied to urban wastes and any organic material can be so treated.

Chemical reduction

Hydrogenation or chemical reduction is achieved by reacting gaseous hydrogen or carbon monoxide with organic materials at temperatures between 300°C and 350°C and at pressures of 2000–4000 pounds per square inch (140–281 kg cm^{-2}). This is technically feasible and generally produces a high fuel-value oil.

Biophotolysis of water

Biophotolysis of water is more speculative and involves the direct formation of hydrogen using the photosynthetic apparatus of, for example, blue–green algae.

The relative efficiencies of some of the six processes described above are indicated in Table 15.7.

TABLE 15.7 The efficiency of some processes for conversion of organic matter into fuel (from Lalor, 1975)

Process	Nature of fuels	Efficiency (% of original heat content in manufactured fuels)
Fermentation	Methane	60% (77% max)
Pyrolysis	Oil	82%
Pyrolysis	Charcoal	
Chemical reduction	Oil	37% (65% expected)

References

Chedd, G. (1975). Cellulose from sunlight. *New Scientist* **65** (939), 572–575.
Clymo, R. S. (1978). A model of peat bog growth. *In* "Production Ecology of British Moors and Montane Grasslands", (Eds O. W. Heal and D. F. Perkins), 187–223. Springer-Verlag, Berlin.
Cousins, W. J. (1975). Preliminary estimates of energy usage in energy farming of trees. Proceedings of Symp on the Potential for Energy Farming in New Zealand. DSIR Information Series No. 117, 77–82.
Downing, C. G. E. (1975). Energy and agricultural biomass production and utilization in Canada. *In* "Energy, Agriculture and Waste Management", (Ed. W. J. Jewel), 261–269. Proc. 1975 Cornell Agricultural Waste Management Conf. Ann Arbor Science Publications, Ann Arbor, Michigan.
Earl, D. E. (1975). "Forest Energy and Economic Development", Clarendon Press, Oxford.
F.A.O. (1971). "Agricultural Commodity Projections 1970–1980", Vol. 1. F.A.O., Rome.
Graham, R. W. (1975). Fuels from crops: renewable and clean. *Mechanical Engineering*, May, 27–31.
Lalor, E. (1975). Photobiological conversion. *In* "Solar Energy for Ireland", Report to National Science Council, Dublin. 40–47.
Leach, G. (1975). "Energy and Food Production", Int. Inst. for Environment and Dev. London and Washington, D.C.
Levitt, J. (1975). Fuel as an agricultural crop. *Energy Conversion* **14**, 93–96.

M.A.F.F. (1975). Energy allowances and feeding systems for ruminants. Technical Bulletin 33. H.M.S.O.

Miller, E. C. (1931). "Plant Physiology", McGraw-Hill, New York.

National Academy of Sciences (1980). Firewood Crops. Shrub and Tree Species for Energy Production. Nat. Academy of Sciences, Washington D.C.

Pimental, D., Hurd, L. E., Bellotti, A. C., Forster, M. J., Oka, I. N., Sholes, O. D. and Whitman, R. J. (1973). Food production and the energy crisis. *Science N.Y.* **182**, 443–449.

Szego, G. C. and Kemp, C. C. (1973). Energy forests and fuel plantations. *Chemtech*. May, 275–284.

Szetela, E. J., Krascella, N. L., Blecher, W. A. (1973). Mariculture investigation: ocean farming and fuel production. Technical Report No. UARL N911599–4 by United Aircraft Research Laboratories, Hartford, Conn. under NSF grant G1–34991. Appended to the Annual Report "Technology for Conversion of Solar Energy to Fuel Gas", No. NSF/RANN/SE/G1–34991/PR/73/4 by the University of Pennsylvania under NSF/RANN Grants G1–29729 and G1–34991.

Tillman, D. A. (1978). "Wood as an Energy Resource", Academic Press, New York and London.

16

PRODUCTION FROM LOWER PLANTS

Lower plants have been used for a very long time, primarily as agents of fermentation, and are of local importance as major items of human diet. Large-scale production, mainly in the form of single-cell protein (S.C.P.), has developed only recently, i.e. in the last ten years or so.

Lower plants have been considered to have two major advantages; (a) very high rates of production per unit of time and of stock and (b) very high protein contents. The first advantage is spectacularly illustrated by the often-repeated claim that, whilst a half-ton (508 kg) bullock takes a day to lay down 1lb (0.45 kg) of protein, half a ton (508 kg) of yeast produces 50 tons (50.8 metric tonnes) in the same period of time — an advantage of 100 000 to one.

The speed at which production can proceed is an aspect of efficiency most clearly demonstrated with lower plants. However, if enough bullocks were gathered together and their starting points staggered in time, any given rate of beef production could be arranged. The difference would then be in the mass of "stock" or capital that was required to sustain such a rate. There are thus interactions between the efficiency of production per unit of time and per unit of biological capital. There is an inevitable delay in production from, for example, large animals and trees, starting from the very beginning and, even when a high rate of production per unit of time has been achieved, a substantial biological capital is then needed to sustain it. The "maintenance cost" is unavoidably high in these circumstances, both in terms of the input of nutrients (including energy) and of all the other resources (such as labour and money) that are required in some relation to the magnitude of this capital.

One measure of rate of production is "protein doubling time" (the time required to double initial weight) (Worgan, 1973) and this is illustrated in Table 16.1.

The protein content of lower plants appears very high (see Table 16.2) and yield per unit area correspondingly so (Table 16.3), but reservations have to be made about the biological value of this protein. Some of the protein may be insoluble cell wall fractions that are only digested by humans to a very limited extent (Tannenbaum *et al.*, 1966) and some, such as yeasts, contain very high levels of nucleic acids (Mateles and Tannenbaum, 1968).

TABLE 16.1   Rate of conversion of carbohydrate to protein (doubling time) by micro-organisms (after Worgan, 1973)

	Doubling time (hours)
Yeasts	
Large-scale batch production	5
Continuous process bakers' yeast	3–4
Candida utilis	1.7–4.7
Fungi	
Neurospora crassa	2
Fungal mycelium in 3 000 l fermenter	2
Bacteria	
Aerobacter aerogenes	0.8–4.7

TABLE 16.2   The protein content of single-cell organisms (after Heydeman, 1973)

	Crude protein in dry weight (%)
Fungi	20–45
Yeasts	40–60
Algae	30–60
Bacteria	50–75

TABLE 16.3 Protein yield of micro-organisms per unit of land (after Vincent, 1969)

	Protein dry weight $(\text{kg ha}^{-1}\text{yr}^{-1})$
Spirulina platensis	24 304
Chlorella pyrenoidosa	15 680

Two other features distinguish most production systems using lower plants. First, they are generally based on quite different resources from those used in higher plant production. Secondly, they have unusual properties as raw material for subsequent processing; this is especially true of the multicellular mycelium of the *Fungi Imperfecti* which possesses structural properties that are important in manufacture and cooking (Spicer, 1971).

The resources used include some quite new to food production and some old ones used in new ways.

Lignocellulose wastes occur in vast quantities in many countries and can be converted by micro-organisms into animal feed. The main wastes include manure and vegetable waste, urban refuse and industrial wastes, municipal sewage solids and wood wastes. The major treatments have been ensilage with mixed anaerobes, mould growth (e.g. *Trichoderma viride*) on waste paper, the use of aerobic mesophiles on bagasse and aerobic thermophiles (such as *Thermoactinomyces*) on livestock wastes (Bellamy, 1976a,b).

The most important efficiency ratios in these cases are likely to relate to the costs incurred and, in particular, the energy costs of the processes involved. Such considerations are likely to outweigh increases in the amount of nitrogen recovered unless these are very large: in any case, this is taken into account if costs are expressed per unit of protein produced.

Energy costs are likely to be of even greater importance in those processes that use hydrocarbons as substrates (Shacklady, 1969; Howard, 1971). Raw materials include natural gas, normal paraffins and gas oil. It now seems unlikely that the use of any energy source that could be used as a fuel (or even for the production of materials) would be sensibly used as the primary energy source for food production (especially for animal feeds), in spite of the obvious importance of food supply. Using oil to help in using more of the incident solar radiation is a quite different matter.

Use of agricultural wastes by micro-organisms seems very much more likely (Howard, 1971) and a powerful case can be made for the use of fungi

(Gray, 1966; Litchfield, 1968), even on main crops of carbohydrate, such as potatoes, sugar beet, sweet potatoes and cassava. Imperfect fungi have been found capable of synthesizing one unit weight of protein from six to eight unit weights of carbohydrate utilized. Furthermore, there is still room for much further exploration amongst the vast number of species and strains of known imperfect fungi. One of the commonest expressions of efficiency in these cases is the protein produced (g) per 100 g of reducing sugar supplied. A selection of such efficiency ratios from the literature is given in Table 16.4.

TABLE 16.4   Protein produced (g) by micro-organisms per 100 g reducing sugar supplied (after Litchfield, 1968)

Organism	Substrate	Protein efficiency (g per 100 g reducing sugar supplied)
Fungi		
Agaricus blazei	Orange juice	11.0
Cantharellus cibarius	Beet molasses	1.0
Morchella spp.	Cheese whey	8.1

Another way of expressing the same efficiency is as biological efficiency of protein production (*BEP*), from the formula

$$BEP = \frac{W \times R \times (S \times A)}{N \times 100}$$

where W = g carbohydrate to yield 100 g protein; N = net protein utilization (*NUP*); R = protein doubling time (h); S = starch equivalent (*SE*); $A = 1$ unless otherwise specified.

The greater the value of this *BEP* index the lower the efficiency (see Table 16.5 for examples). Many lower plant production processes are based on liquid culture, generally in fresh water, but developments for mass production of marine algae in outdoor cultures have also been proposed (Goldman *et al.*, 1975). Results, using such diatom species as *Phaeodactyluna tricornutum, Amphiprora* spp., *Amphora* sp. and *Nitzchia closterium*, at Woods Hole, Massachusetts, and Ft. Pierce, Florida, have shown that up to 19 g DM m^{-2} per day can be produced at the right dilution rate. This was on mixtures of "secondarily treated domestic wastewater" and seawater in fully continuous-flow cultures.

The production levels can be compared with yields for freshwater algae and for marine mass cultures (see Table 16.6): they may also be compared

TABLE 16.5 Examples of the biological efficiency of protein production by micro-organisms (after Worgan, 1973)

	BEP index
Yeasts	
Candida utilis	
Batch culture	5.3
Continuous culture	3.2
Fungi	
Fusarium semitectum	3.1
Bacteria	
Escherichia coli	2.8

TABLE 16.6 Production level of marine algae cultures compared with freshwater algae cultures (after Goldman *et al.*, 1975)

	Dry weight yields ($g\ m^{-2}$ per day)
Freshwater algae (Europe)	10–20
Marine mass culture	5–12
Maximum yield (Ryther, 1959)	27.0

with Ryther's (1959) estimate, based on sunlight availability, that the maximum sustainable yield in a body of water is about 27 g DM m^{-2} per day. Clearly this "ceiling" value can be used to assess the efficiency of any particular production process and to explore the potential value of projected developments.

Production values of this order compare very favourably with fast-growing crops and annual production may be vastly better. This is because, although the maximum rates of dry matter production may be as high for higher plants (*c.* 50 g m^{-2} per day; Verduin, 1953) as for algae (*c.* 50.4 g m^{-2} per day; Hindak and Pribl, 1968), the latter need far less mass, since all cells photosynthesize (Heydeman, 1973). They also have a higher protein content, and potential annual production of *Chlorella* in 5% CO_2 has been suggested to be 60–90 × 10^3 kg DM ha^{-1}, containing 50% protein (Dean,

1958), compared with yields of soyabeans with 33% protein, of the order of 1.5×10^3 kg DM ha^{-1} yr^{-1}.

Furthermore, some algae are easy to harvest and have been used as human food for centuries: *Spirulina maxima* in Chad is an example.

Since the emphasis is so often on protein production, the efficiency with which nitrogen is utilized matters and it is surprising that relatively little attention has been given, in the past, to nitrogen-fixing micro-organisms for the direct production of human food (see Table 16.7 for yields of these). Of course, where the function of micro-organisms is to recover nitrogen from wastes, this is not relevant, but as an independent production process the ability to fix atmospheric nitrogen coupled with high protein content and high specific growth rates suggests an extremely efficient combination. Since labour or support energy usage might also be high, it may be worth considering on how small a scale such systems could be viable, using the labour supply available to poor and relatively landless, hungry households. Since the most developed agricultural systems have been devised, in general, to use as little labour as possible, it is sometimes forgotten that in many parts of the world labour is plentiful and that it is one of the few resources that actually increases with increase in population size.

TABLE 16.7 Yields of edible nitrogen-fixing organisms (from Mishustin and Shil'nikova, 1971)

	Nitrogen per g glucose consumed (mg)
Aerobacter aerogenes	
in anaerobic conditions	2.5–4.2
in aerobic conditions	3.6–3.8

A plentiful labour supply could transform the outlook for many potential agricultural systems and make it possible to visualize many small production units rather than a few huge, automated complexes. The prospects for such processes must often be calculated in terms of economic efficiency and labour put into the calculation at a fair price, but this is not the right framework for many hungry people. Furthermore, it is necessary to distinguish carefully between labour used directly in production processes (which is labelled as such) and labour used in producing machinery (which is hidden in the machinery costs). It may also become important to distinguish between the use of tools by men and the substitution of other sources of power for the men. Tools will generally be sensible as a means of saving energy but shortage of fuel may cause a review of the relative merits of labour and oil as sources of power.

The relevance of this discussion at this point has to do with the variety of ways in which micro-organisms could be used. They already perform an enormous range of functions, often largely unrecognized (Dixon, 1976), but, because of their small size, they can be embodied in processes of almost any scale.

It would be a pity if their remarkable qualities were unused because the energy requirements of large-scale production proved uneconomic.

References

Bellamy, W. D. (1976a). Production of single-cell protein for animal feed from lignocellulose wastes. *World Animal Review* No. 18, 39–42.

Bellamy, W. D. (1976b). Cellulose and lignocellulose digestion by thermophilic *Actinomyces* for single cell protein production. Paper presented at the 33rd General Meeting of the Society for Industrial Microbiology Aug. 1976.

Dean, R. F. A. (1958). Use of processed plant proteins as human food. *In* "Processed Plant Protein Foodstuffs", (Ed. A. M. Altschul), 205–47. Academic Press, New York and London.

Dixon, B. (1976). "Invisible Allies", Temple Smith, London.

Goldman, J. C., Ryther, J. H. and Williams, L. D. (1975). Mass production of marine algae in outdoor cultures. *Nature (London)* 254(5501), 594–5.

Gray, W. D. (1966). Fungi and world protein supply. *Adv. Chem.* 57, 261–268.

Heydeman, M. T. (1973). Aspects of protein production by unicellular organisms. *In* "The Biological Efficiency of Protein Production", (Ed. J. G. W. Jones). Cambridge University Press, London.

Hindak, F. and Pribl, S. (1968). Chemical composition, protein digestibility and heat of combustion of filamentous green algae. *Biologia Plantarum* 10(3), 234–44.

Howard, J. (1971). New proteins: animal, vegetable, mineral. *New Scientist and Science J.* 49(740), 438–439.

Litchfield, J. H. (1968). The production of fungi. *In* "Single-cell Protein", (Eds R. I. Mateles and S. R. Tannenbaum), MIT Press, Cambridge, Massachusetts.

Mateles, R. I. and Tannenbaum, S. R. (Ed) (1968). "Single-cell Protein", MIT Press, Cambridge, MA.

Mishustin, E. N. and Shil'nikova, V. K. (1971). "Biological Fixation of Atmospheric Nitrogen", Macmillan, London.

Ryther, J. H. (1959). Potential productivity of the sea. *Science (N.Y.)* 130, 602–8.

Shacklady, C. A. (1969). Proteins from paraffins. *New Scientist* 43(668), 5–7.

Spicer, A. (1971). Synthetic proteins for human and animal consumption. *Vet. Rec.* 89, 482–486.

Tannenbaum, S. R., Mateles, R. I. and Capco, G. R. (1966). Processing of bacteria for production of protein concentrates. *Adv. Chem.* 57, 254–260.

Verduin, J. (1953). A table of photosynthetic rates under various conditions. *Am. J. Botany* 40, 675–679.

Vincent, W. A. (1969). Algae for food and feed. *Process Biochemistry* 4(6), 45–47.

Worgan, J. T. (1973). Protein production by micro-organisms from carbohydrate substrates. *In* "The Biological Efficiency of Protein Production", (Ed. J. G. W. Jones). Cambridge University Press, London.

Part III
Efficiency in Animal Production

17

ADAPTATION TO THE ENVIRONMENT

There are many ways in which animals may be adapted to their environments and many ways in which agricultural practice can influence the
environment. Since animals can be concentrated more easily than plants, it
is often easier to provide some environmental control for them than for
crops. High value crops can be "housed" just as well as animals but the
limited areas in which this is economic reduce its application to large-scale
crop production. In the case of animals, they can be brought together all the
time or just for critical periods, so the additional implications, of providing
feed and removing waste products, are extremely varied.

It is quite likely that the efficiency of one process may be increased by
housing, whereas overall (or another) resource use efficiency may be decreased. For example, efficiency of feed use may be increased but efficiency
of support energy use decreased by environmental control (Turner, 1974).
Much depends, of course, on the degree of environmental control that is
required to improve performace or, indeed, to make it possible to keep a
particular kind of animal at all. It may seem excessive to contemplate
providing the heat needed to enable tropical animals to be kept in cold or
even temperate regions, yet it would be taken for granted that tanks or
ponds must be provided if fish are to be farmed in areas without natural
water. On the other hand, environmental control is expensive and strong
arguments are advanced (e.g. Crawford, 1974) for the proposition that the
wild animal species of the semi-arid lands in Africa should be domesticated,
rather than trying to insert temperate animals and practices from northern
Europe.

The provision of environmental control or protection is thus a complicated matter and varies from houses of one sort or another to various types of
coat. All this relates to the climate but other features of the environment
may be of equal importance in influencing different aspects of efficiency.
Notable amongst these are disease incidence, feed and water supply, soil
type and topography.

Soil type can influence what animals are, or should be, kept and animals may be adapted structurally to sand (e.g. camels), mud (e.g. buffaloes) or boggy land (e.g. geese). Soil type also influences important interactions between grazing animals and the amount of herbage grown and harvested.

Topography similarly matters to animals; goats are physically able to browse where cattle cannot go, for example.

Disease incidence may be of immense importance. The presence of trypanosomiasis and its vector, the tsetse fly (*Glossina morsitans*), currently renders vast tracts of tropical Africa totally unsuitable for most breeds of cattle. There are some immunotolerant breeds, however, such as the N'Dama and West African Shorthorns, thus illustrating the possibility of adaptation (Jahnke, 1976), but it is also possible to alter the vegetation (partly by grazing with goats) in such a way that it becomes unsuitable for tsetse flies. The environment can be altered in many ways in order to change disease incidence: liver flukes (*Fasciola hepatica*) can be controlled by drainage to eliminate the snails (*Limnaea truncatula*) that serve as secondary hosts, for example. Housing and treatment of animals can sometimes control both external and internal parasites, and treatment of the land or the vegetation may control other forms of disease-producing organisms.

Adaptation by the animal can take several forms. Resistance or immunity may be of genetic, acquired or, possibly, nutritional origin, and behavioural responses may be significant in determining the extent to which animals are exposed to disease or not.

The influence of adaptation on efficiency is thus a very large subject and it is proposed to restrict this discussion to four examples, relating to (a) climate, (b) topography, (c) water supply and (d) feed supply.

Climate

Individual animals may be adapted to climatic conditions by virtue of structural, physiological or behavioural attributes (Smith, 1975; Ingram and Mount, 1975; Thompson, 1976; Robertshaw and Finch, 1976). Warm-blooded animals are generally insulated against heat loss, by coat (hair, wool or feathers) or fat cover, and provided with cooling mechanisms, generally involving sweating, to allow rapid responses to temperatures outside the thermoneutral zone. The need to do this depends also on the extent of heat production by the animal body and this is governed not only by characteristic metabolic rates but by level of performance and activity. A cow on a low plane of nutrition may need to conserve heat in a cold climate, but the same cow at the same ambient temperature may need to dissipate heat if she is on a high plane of nutrition and producing a high milk yield (Webster, 1976).

Rumen fermentation produces heat and the plane of nutrition influences the quantity produced.

Other climatic features, such as rain and wind, often act together to modify the influence of temperature. The rate of heat loss is increased by wind and the insulation of the coat, particularly, is greatly reduced by rain. Some animals, such as mountain sheep, have coats with woolly components for insulation underneath hairy outer layers that shed rain. Subcutaneous fat can be a very effective insulatory layer, as in the pig.

The same kind of coat, for example in thick-woolled sheep, that serves for insulation may also protect against excessive insolation. Once a sheep has about half an inch of fleece it is almost completely insulated, so additional wool does not necessarily add to its heat load. In tropical conditions with intense sunshine, however, the temperature of the sheep may be substantially lower than that of the wool surface, with the fleece acting as a shade.

Shade, of course, may be another reason for the provision of houses, or trees, and overheating may also be prevented by spraying with water.

The possible effect of climate on efficiency can be neatly illustrated by reference to the effect of ambient temperature on egg production of the domestic hen (Fig. 17.1) and the consequent effect on efficiency of feed conversion (Table 17.1). In general, low temperatures (i.e. low relative to requirement) result in non-productive use of the feed eaten and high temperatures result in low performance because of low feed intake, unless

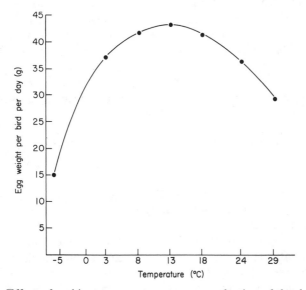

FIG. 17.1   Effect of ambient temperature on egg production of the domestic hen (after Payne, 1966)

TABLE 17.1   Effect of ambient temperature on egg production of the domestic hen and the consequent effect on efficiency of feed conversion (after Payne, 1966)

Air temperature (°C)	Egg production (g egg per bird per day)	Daily feed intake (g per bird)	g egg per g feed
−5	15.0	186	0.08
3	37.0	158	0.23
8	41.7	150	0.28
13	43.5	141	0.31
18	41.3	132	0.31
24	36.6	122	0.30
29	29.3	113	0.26

nutrient intake can be maintained by adjusting the nutrient concentration in the diet.

An individual may respond to temperature changes by seeking warmer or cooler places, by physiological responses or by curling up to reduce the area of body surface exposed or stretching out to expose more. Animals may also huddle together in rain or wind (and to escape the attention of flies, incidentally). Young animals may often be protected by their dams in this way.

Population dynamics are also of some significance in influencing which animals are exposed, in what physiological state, at what season of the year. This is in some ways even more true of agricultural animals than of wild animals. Lambs are most commonly born in the spring, for example, and are frequently slaughtered before the next winter, so that very young sheep are not exposed to winter weather at all. In cattle, growth to normal slaughter weights takes much longer (one to two years), and calving is usually rather less markedly seasonal.

Topography

Topographical features of the environment interact with other aspects, such as climate, water and nutrient supply but, in addition, contribute some special influences.

Amongst these are effects on the availability of shelter, which can modify the local climate of animals, and similar effects on crop growth, including the provision of forage for animals.

A rather special and quite important effect, however, is that on the energy cost of animal movement.

It is generally agreed (see Riberio *et al.*, 1977) that, for cattle moving at walking speed, the energy cost of horizontal locomotion is about 2 J kg^{-1} m^{-1}, independent of body weight or plane of nutrition. Values quoted for other species (in J kg^{-1} m^{-1}) are red deer 2.6, dog 2.5, Man 2.3, sheep 2.3 and horse 1.6.

A similar constancy appears to exist for vertical ascent (26.5 J kg^{-1} per vertical metre for cattle, 26.6 for sheep and 21.5 for red deer) but at a much higher cost. The energy cost of downhill locomotion is not known.

The energy cost of walking has been estimated (Ribeiro *et al.*, 1977) as increasing ME requirement by between 4% and 24%, according to circumstances.

Water supply

An animal's requirement for water varies with climatic conditions, activity and physiological state (Webster and Wilson, 1966; Macfarlane *et al.*, 1971; Hohenboken and Kistner, 1976). Some, especially small mammals on dry diets, can survive for only a short time (a few days or less) without drinking: others, notably the camel, can go for up to 20 days without taking in water (even in food).

Physiological mechanisms exist for reducing water loss, by concentration of urine, for example, for metabolizing water, in the utilization of reserve fat, and for recycling water within the body. Normally, recycling of water is substantial in many animals. For instance, the daily output of saliva by a cow on dry feed may be of the order of 150 l (Church, 1976), most of which is reabsorbed from the alimentary tract.

Since animals differ greatly in their needs (see Table 17.2), some may tolerate arid environments much better than others (Table 17.3). Even when animals can survive, performance may be reduced by water shortage: milk production may be lowered if cows lack drinking water, for example, and in other cases the intake of dry food will be reduced. The effects of water shortage on efficiency, therefore, will generally be to lower it by reducing output, directly or by reduced feed intake.

Feed supply

This is an important topic for several reasons. Agriculturally, feed is normally the biggest single cost input to animal production, so feed conversion efficiency is extremely important economically. Biologically, output per unit of land depends upon the use made of the feed grown on it: thus carrying capacity is greatly influenced by feed conversion efficiency and is often

TABLE 17.2   Water requirements for different animals (from Spedding and Hoxey, 1975)

Species	Liveweight (kg)	Environment	Water (l/100 kg liveweight/day)	Sources[a]
Beef cattle	400	15–21°C	7	(1)
Bos taurus		Arid subtropics (summer)	16.1	(2)
Bos indicus		Equatorial arid (July)	7.5	(2)
Musk oxen	365	Arctic (April)	3.5	(2)
Sheep	40	15–21°C	9	(1)
Haired sheep	18–25	Simulated desert	3.8	(3)
Sheep (Ogaden fat tail)		Arid subtropics (summer)	10.2	(2)
Sheep (Corriedale)		Arctic (April)	6.2	(2)
Goats		Arctic (April)	5.2	(2)
Goats	15–20	Simulated desert	4.2	(3)
Pigs	15	Normal housing	7	(4)
Pigs	90		6	
Horses	600	—	4.1–14.3	(5,6)
Reindeer		Arctic (April)	12.8	(2)
Camel		Arid subtropics (summer)	8.2	(2)
Rabbits	2.5	Housed: dry diet	6	(7)
Hens	2–3	Housed: dry diet	7.35	(8)

[a]Sources: (1) A.R.C. (1965); (2) Macfarlane *et al.* (1971); (3) Maloiy and Taylor (1971); (4) A.R.C. (1966); (5) Rossier (1973); (6) Olsson (1969); (7) Kennaway (1943); (8) Tyler (1958)

TABLE 17.3   Animals which will tolerate arid environments

Cold deserts
Llamas	*Lama* spp.
Bactrian camels	*Camelus bactrianus*
Goats	*Capra hircus*
Sheep	*Ovis aries*

Hot deserts
Camel	*Camelus dromedarius*
Eland	*Taurotragus oryx*
Zebu cattle	*Bos indicus*
Sheep e.g. Blackhead Persian and Somali	
Goats	

Source: Macfarlane (1964)

dominated by the supply of feed, both in total and in terms of seasonal patterns. Marked seasonality of yield is a characteristic of both temperate (see Spedding and Diekmahns, 1972) and tropical (Osbourn, 1975) pastures.

Adaptation of animals to feed supply takes many forms. For the collection of feed, animals are modified in innumerable ways. The length and structure of their legs influences their capacity to collect feed from tall vegetation or herbage sparsely distributed over large areas. Mouth, tongue, lips and teeth are adapted to different methods of prehension and severing of vegetation. Related to methods of harvesting that are required or advantageous in the wild, many herbivores have capacious stomach compartments for use as temporary stores of grazed herbage. These, such as the rumen and caecum, have often become fermentation chambers of enormous significance to agriculture.

Indeed, it is the ability of an animal's ailmentary tract to deal with different feeds that affects efficiency more than any other single adaptation, apart from those that virtually determine survival.

The nature and structure of the alimentary tract determines the quantities that can be eaten, the time for which food can remain within it, the digestive processes to which food can be subjected and the way in which digested nutrients are absorbed.

The quantity that an animal can eat varies with many factors. Nevertheless, there are large differences between species in the maximum amount of dry matter that can be consumed per day (Table 17.4). Similarly, rate of passage through the gut varies with the nature of the diet and with the amount eaten, but there are characteristic durations for normal conditions (Table 17.5).

Although many animals are specialized to digest particular diets, and in some cases this specialization is quite extreme, most agricultural animals

TABLE 17.4 Maximum amount of dry matter that can be consumed per day by different animals

Animal	Liveweight (kg)	Total intake of DM (kg per day)	Intake of DM (g per kg Lwt per day)
Dairy cows (lactating)	500	17.5	35
Sow (lactating)	225	6.6	29
Ewe (lactating)	70	2.83	40
Laying hen	2.7	0.136	50

Sources include: Greenhalgh (1969); Hill (1969); A.R.C. (1966); A.R.C. (1965)

TABLE 17.5 Rate of passage of feed through different animals under normal conditions

	Retention time[a] (h)	Sources[b]
Cattle	70–90	(1)
Sheep	60–110	(2)
Goats	32–65	(3)
Pigs	48–72	(4)
Geese	2	(5)
Grass carp	7.5[c]	(6)
Snail (Helix pomatia)	6	(7)

[a]This varies with the nature of the diet and the level of intake. It may be expressed in various ways: here it refers to the time by which about 90% of the faeces has been excreted.
[b]Sources: (1) Balch and Campling (1965); (2) Blaxter et al. (1956); (3) Castle (1956); (4) Parker and Clawson (1967); (5) Mattocks (1971); (6) Hickling (1966); (7) Turček 1970
[c]at 27°C

can, in fact, make some use of quite a wide range of feedstuffs. The extent of the use made thus varies widely and results in great variation in feed conversion efficiency.

Digestibility is usually defined in one of the following ways:

"True" digestibility = % of the feed that is digested and absorbed;

$$\text{"Apparent" digestibility} = \frac{\text{Amount eaten} - \text{quantity of faeces}}{\text{Amount eaten}} \times 100$$

all expressed in the same terms (e.g. dry matter, organic matter, energy or protein) and measured for the same period.

Digestibility is thus an efficiency ratio: the difference between the two expressions largely reflects the fact that faeces contain dead bacteria and sloughed-off gut wall tissue, in addition to the undigested parts of the feed. The faecal output therefore overestimates the amount that has not been digested; nevertheless, the apparent digestibility may be used as an approximation to digestive efficiency and, as it is more easily measured, most data are available in this form.

Digestibility may vary with the age of the animal, the level of feed intake and even the level of internal parasitism. There is not just one figure that will be valid for all animals, even of animals that are of the same species and breed and are similar in physiological state, eating the same diet.

Even so, the usefulness of the digestibility concept is that it can characterize a feed in relation to a particular kind of animal. Furthermore, differences between animal species can be detected, on the same feed, indicating their relative abilities to digest that feed. Examples of these differences are shown in Table 17.6, for a wide range of animal species on some contrasting diets. An illustration of the way in which digestive efficiency may vary with level of feed intake is given in Fig. 17.2.

TABLE 17.6   Digestibility of a range of feeds by different animal species

(a) Digestion coefficients (after Vanschoubroek and Cloet, 1968)

	Rabbit		Pig		Sheep		Cattle	
	CP	CF	CP	CF	CP	CF	CP	CF
Peas	63.9	−0.5	81.3	15.4	82.5	54.8	85.9	56.3
Barley	32.7	−4.1	71.1	30.0	59.7	48.3	56.1	25.0
Maize	24.8	–11.8	70.0	41.1	48.7	−48.1	52.0	3.6
Soyabeans	74.2	50.6	79.5	55.3	90.9	84.7	91.6	55.7

(b) Digestibility (after Hintz et al., 1973)

	New world camels		Sheep		Ponies	
	CP	ADF	CP	ADF	CP	ADF
Alfalfa pellets	74.7	61.0	69.7	50.5	65.7	46.7
Hay-grain pellets	69.2	38.7	63.1	34.0	68.8	29.3

(c) Digestibility (after Slade and Hintz, 1969)

	Horse		Pony		Rabbit		Guinea-pig	
	CP	CF	CP	CF	CP	CF	CP	CF
Alfalfa pellets	74.0	34.7	76.2	38.1	73.7	16.2	69.0	38.2
Alfalfa-grain pellets	77.3	38.6	69.6	40.9	73.2	18.1	—	—

CP, crude protein; CF, crude fibre; ADF, acid detergent fibre

It will be obvious from Table 17.6 that some species are much better adapted to certain diets than are others. Before pursuing this adaptation further, however, it should also be noted that feeds do not have a constant digestibility even for the same animal. Since the extent to which an animal can digest plant material depends upon its structure and chemical composition, and these change during plant growth, the digestibility of a plant species also changes as the plant grows (Fig. 17.3), as well as varying from one plant part to another (see Fig. 17.4). The result of the last point is that if an animal feeds selectively on a plant, as in much grazing and browsing, the digestibility of what is eaten may differ markedly from the mean for the plant as a whole.

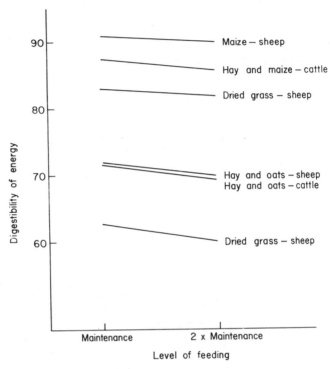

FIG. 17.2   The apparent digestibility of feed energy by sheep and cattle as affected by feeding level (after Blaxter, 1961)

Adaptations of the alimentary tract to particular diets takes several forms. Where mouth parts do not include teeth, as in birds, a grinding mechanism is provided by a specially adapted muscular gizzard (Fig. 17.5). The need for comminution of the feed depends to some extent on whether enzymes are present that can break down cell walls or whether this has to be accomplished physically to release the cell contents.

Where feeding may be fairly continuous, as in herbivorous fish, for

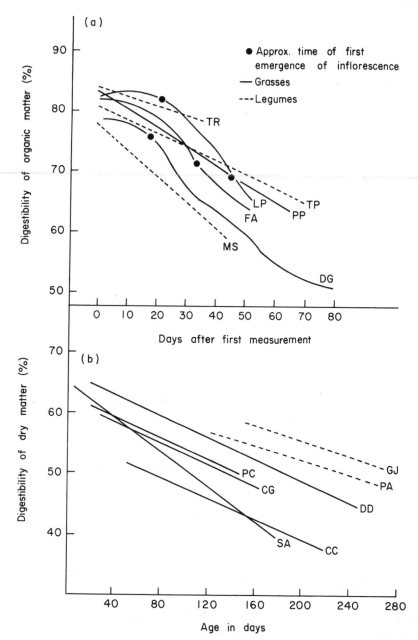

FIG. 17.3   Changes in the digestibility of temperate and tropical grasses and legumes with increasing age and maturity of the herbage (from Corbett, 1969). (a) Temperate species: Spring growths; dates of first measurements varied between species and

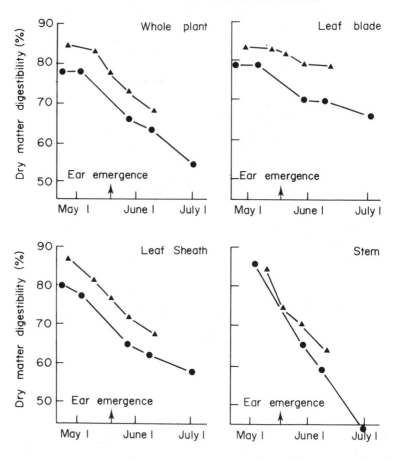

FIG. 17.4   In-vitro digestibility (per cent of dry matter) of fractions of S24 ryegrass (▲) and S37 cocksfoot (●) (first cuts taken in Spring 1959) (from Terry and Tilley, 1964)

FIG. 17.3 (*cont.*)

between localities. LP, *Lolium perenne* var. S23 (ryegrass); PP, *Phleum pratense* var. S48 (timothy); FA, *Festuca arundinacea* var. S170 (tall fescue); DG, *Dactylis glomerata* var. S37 (cocksfoot or orchard grass); TR, *Trifolium repens* var. S100 (white clover; TP, *Trifolium pratense* var. Ultuna (red clover); MS, *Medicago sativa* var. Dupuits (lucerne or alfalfa). (b) Tropical species: SA, *Sorghum almum*; CC, *Cenchrus ciliaris* var. Molopo (Buffel grass); CG, *Chloris gayana* var. Callide (Rhodes grass); PC, *Pennisetum clandestinum* (Kikuyu grass); DD, *Digitaria decumbens* (Pangola grass); PA; *Phaseolus atropurpureus* var. Siratro; GJ, *Glycine javanica* var. Cooper.

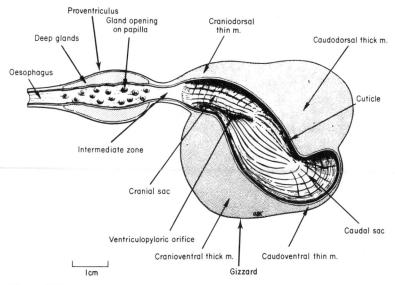

FIG. 17.5   Interior stomach (gizzard) of the domestic fowl (from King and McLelland, 1979)

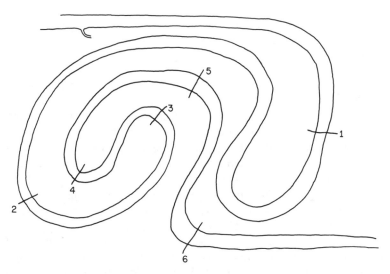

FIG. 17.6   Empty gut of *C. idella* (grass carp) as dissected out, with the approximate positions of the six ligaments (from Hickling, 1966)

example, the gut may be short, with no large organs (Fig. 17.6). Simple-stomached animals have a substantial organ in which primary digestion occurs and the total length of the gut may be considerable (Table 17.7).

TABLE 17.7   Length of intestines of different species in metres

	Pig	Ox	Sheep Goat	Horse	Herbivorous fish Grass[a] carp	Others[b]
Small intestine	16.0–21.0	27.0–49.0	18.0–35.0	19.0–30.0		
Duodenum	0.7–0.95	0.9–1.2	0.6–1.2	1.0–1.5		
Jejunum ⎫				17.0–28.0		
⎬	15.0–20.0	26.0–48.0	17.5–34.0			
Ileum ⎭				0.7–0.8		
Large intestine	3.5–6.0	6.5–14.0	4.0–8.0	6.0–9.0		
Caecum	0.3–0.4	0.5–0.7	0.25–0.42	0.8–1.3		
Colon and rectum	3.0–5.8	6.0–13.0	3.5–7.5	5.5–8.0		
Total length of gut	20.0–27.0	33.0–63.0	22.0–43.0	25.0–39.0	2.25	
Total body length	1.4–2.2	2.0–2.4	up to 1.2	2.0–2.8	up to 1.0	
$\dfrac{\text{Total length of gut}}{\text{Total body length}}$	14.3–12.3	16.5–26.3	35.8	12.5–13.9	2.25	12–17

[a]The grass carp has a short alimentary tract, typical of fish that feed on higher plants
[b]Longer tracts are associated with feeding on microscopic plants

Sources: Nickel et al. (1973); Kapoor et al. (1975); Hickling (1966)

The ruminants, have, in addition, four chambers to the stomach, including a voluminous rumen (see Fig. 17.7) in which fermentation by micro-organisms can occur. Similar processes take place in the colon of the horse and the caecum of the rabbit (Fig. 17.8). Such organs are not functional at birth but develop as the animal is weaned from milk on to solid feed. Even then, the nature and rate of development varies with the nature of the feed, so there is a degree to which individual animals are adapted to particular diets. Indeed, this is to some extent a general finding and abrupt changes in diet often lead to digestive problems.

Simple-stomached animals cannot digest cellulose and can therefore make relatively little use of fibrous diets. Their feed conversion efficiency on such diets is bound to be low. Ruminants naturally show much higher efficiencies than do simple-stomached animals on fibrous material but they can also live on concentrated feeds. Over recent years, there has been some tendency to feed ruminants on concentrates (e.g. barley beef), in order to increase the efficiency with which feed dry matter, energy or organic matter is converted to animal product. The increase in efficiency is due to higher

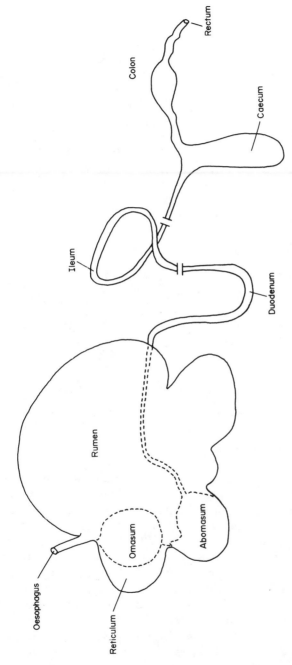

FIG. 17.7   The digestive tract of a ruminant (bovine) (from Maynard and Loosli, 1965)

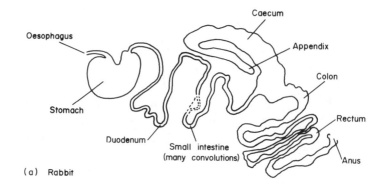

(a) Rabbit

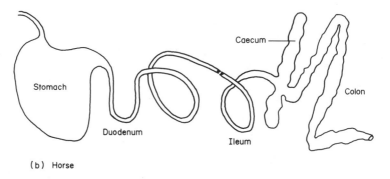

(b) Horse

FIG. 17.8   The digestive tracts of (a) rabbit and (b) horse (horse from Maynard and Loosli, 1965).

performance (e.g. growth rate) on the concentrated feed: economic efficiency depends on the relative costs of the feeds and the initial animal (e.g. the purchased calf) and the price paid for the product. However, it is clear that ruminants so used are no longer fulfilling the purpose of utilizing fibrous feeds and obviously the efficiency with which such feeds are used is not improved by feeding animals on something else, although there are many cases where a mixed diet can be beneficial not only in (a) using a fibrous material but also in (b) obtaining an adequate level of performance. This may be because the quality of the total ration is raised or because the animal is able to eat a greater quantity per day.

Where waste or by-products exist, it may be quite obvious what kind of animal is required and what it is required to eat, since this is one way of utilizing such materials.

If a particular animal product is required, however, it may not be immediately apparent which feed resources should be produced as a basis for the process and it has to be remembered that the rate of animal performance (such as growth rate, milk yield or egg output) may influence the efficiency with which other resources (such as invested capital, time or labour) are used.

References

A.R.C. (1965). "The Nutrient Requirements of Farm Livestock. 2. Ruminants", H.M.S.O., London.
A.R.C. (1966). "The Nutrient Requirements of Farm Livestock. 3. Pigs", H.M.S.O., London.
Balch, C. C. and Campling, R. C. (1965). Rate of passage of digesta through the ruminant digestive tract. In "Physiology of Digestion in the Ruminant", (Ed. R. W. Dougherty), Butterworths, London.
Blaxter, K. L. (1961). The utilization of the energy of food by ruminants. Symp. on Energy Metabolism 2. E.A.A.P. Pub. No. 10, Rome.
Blaxter, K. L., Graham, N. McC. and Wainman, F. W. (1956). Some observations on the digestibility of food by sheep and on related problems. *Br. J. Nutr.* **10**, 69–91.
Castle, E. J. (1956). The rate of passage of foodstuffs through the alimentary tract of the goat. *Br. J. Nutr.* **10**, 15–23.
Church, D. C. (1976). "Digestive Physiology and Nutrition of Ruminants. Vol. 1: Digestive Physiology", 2nd edn, D. C. Church, Corvallis, Oregon.
Corbett, J. L. (1969). The nutritional value of grassland herbage. In "The Nutrition of Animals of Agricultural Importance", Part 2, (Ed. D. Cuthbertson). Pergamon, Oxford.
Crawford, M. A. (1974). The case for new domestic animals. *Oryx* **XII(3)**, 351–360.
Greenhalgh, J. F. D. (1969). Nutrition of the dairy cow. In "Nutrition of Animals of Agricultural Importance", Part 2. (Ed. D. Cuthbertson). 717–770. Pergamon, Oxford.
Hickling, C. F. (1966). On the feeding process in the White Amur, *Ctenopharyngodon idella. J. Zool.* **148**, 408–419.
Hill, F. W. (1969). Poultry nutrition and nutrient requirements. In "Nutrition of Animals of Agricultural Importance", Part 2 (Ed. D. Cuthbertson). 1137–1179. Pergamon, Oxford.
Hintz, H. F., Schryver, H. F. and Halbert, M. (1973). A note on the comparison of digestion by New World camels, sheep and ponies. *Anim. Prod.* **16**, 303–305.
Hohenboken, W. and Kistner, T. P. (1976). Summer water consumption, body temperature and respiration rate in lambs. *Can. J. Anim. Sci.* **56**, 739–744.
Ingram, D. L. and Mount, L. E. (1975). "Man and Animals in Hot Environments", Springer-Verlag, Berlin.
Jahnke, H. E. (1976). Tsetse flies and livestock development in East Africa. A study in environmental economics. Afrika-Studien No. 87, 1–1 80.
Kapoor, B. G., Smit, H. and Verighina, I. A. (1975). The alimentary canal and digestion in Teleosts. *Adv. Mar. Biol.* **13**, 109–239.

Kennaway, E. L. (1943). The supply of water to rabbits and guinea-pigs. *Br. Med. J.* **1**, 760.

King, A. S. and McLelland, J. (1979). "Form and Function in Birds" Vol. 1. Academic Press, London and New York.

Macfarlane, W. V. (1964). Terrestrial animals in dry heat: ungulates. *In* "Handbook of Physiology. Section 4: Adaptation to the Environment", (Ed. D. B. Dill), American Physiological Society, Washington, D.C.

Macfarlane, W. V., Howard, B., Haines, H., Kennedy, P. J. and Sharpe, C. M. (1971). Hierarchy of water and energy turnover of desert mammals. *Nature (London)* **234**, 483–484.

Maloiy, G. M. O. and Taylor, C. R. (1971). Water requirements of African goats and haired-sheep. *J. Agric. Sci., Camb.* **77**, 203–208.

Mattocks, J. G. (1971). Goose feeding and cellulose digestion. *Wildfowl* **22**, 107–113.

Maynard, L. A. and Loosli, J. K. (1965). "Animal Nutrition", 6th edn, McGraw Hill, New York.

Nickel, R. Schummer, A. and Seiferle, E. (1973) "The Viscera of the Domestic Mammals", Translation and revision by W. O. Sack. Verlag Paul Parey, Berlin.

Olsson, N. O. (1969). The nutrition of the horse. *In* "Nutrition of Animals of Agricultural Importance", Part 2, (Ed. D. Cuthbertson) 921–960. Pergamon, Oxford.

Osbourn, D. F. (1975). Beef production from improved pastures in the tropics. *Wld Rev. Anim. Prod.* **XI(4)**, 23–31.

Parker, J. W. and Clawson, A. J. (1967). Influence of level of total feed intake on digestibility, rate of passage and energetic efficiency of reproduction in swine. *J. Anim. Sci.* **26**, 485–489.

Payne, C. G. (1966). Environmental temperature and egg production. *In* "Physiology of the Domestic Fowl", (Eds C. Horton-Smith and E. C. Amoroso). Oliver and Boyd, Edinburgh and London.

Ribeiro, J. M. De C. R., Brockway, J. M. and Webster, A. J. F. (1977). A note on the energy cost of walking in cattle. *Anim. Prod.* **25**, 107–110.

Robertshaw, D. and Finch, V. (1976). The effects of climate on the productivity of beef cattle. *In* "Beef Cattle Production in Developing Countries", (Ed. A. J. Smith). University of Edinburgh Press, Edinburgh.

Rossier, E. (1973). Personal communication.

Slade, L. M. and Hintz, H. F. (1969). Comparison of digestion in horses, ponies, rabbits and guinea pigs. *J. Anim. Sci.* **28(6)**, 842–843.

Smith, A. J. (1975). Beef production in developing countries. *Wld Rev. Anim. Prod.* **XI(1)**, 38–43.

Spedding, C. R. W. and Diekmahns, E. C. (editors) (1972). "Grasses and Legumes in British Agriculture". Commonwealth Agricultural Bureaux, Farnham Royal.

Spedding, C. R. W. and Hoxey, A. M. (1975). The potential for conventional meat animals. *In* "Meat", (Eds D. J. A. Cole and R. A. Lawrie). Butterworths, London.

Terry, R. A. and Tilley, J. M. A. (1964). The digestibility of the leaves and stems of perennial ryegrass, cocksfoot, timothy, tall fescue, lucerne and sainfoin, as measured by an in-vitro procedure. *J. Br. Grassld Soc.* **19(4)**, 363–372.

Thompson, G. E. (1976). Principles of climate physiology. *In* "Beef Cattle Production in Developing Countries", (Ed. A. J. Smith). University of Edinburgh Press, Edinburgh.

Turček, F. J. (1970). Studies on the ecology and production of the Roman snail *Helix pomatia* L. *Biológia (Bratislava)* **25(2)**, 103–108.

Turner, H. G. (1974). The tropical adaptation of beef cattle — an Australian study. *Wld Anim. Rev.* **13**, 16–21.

Tyler, C. (1958). Some water and dry-matter relationships in the food and droppings of laying hens. *J. Agric. Sci., Camb.* **51**, 237–247.

Vanschoubroek, F. and Cloet, G. (1968). The feeding value of concentrates in the rabbit, *Wld Rev. Anim. Prod.* **IV(16)**, 70–76.

Webster, A. J. F. (1976). The influence of the climatic environment on metabolism in cattle. *In* "Principles of Cattle Production", (Eds H. Swan and W. H. Broster), Butterworths, London.

Webster, C. C. and Wilson, P. N. (1966). "Agriculture in the Tropics", Longmans, London.

18

THE SUPPLY OF NUTRIENTS

The nutrient requirement of an animal (A.R.C. 1965, 1966, 1975; N.A.S. 1963, 1968) varies greatly with its species, size, breed, sex and physiological state and with the environment.

If an animal receives less than it requires, it may reduce its performance, or its activity, and remain perfectly healthy, or it may suffer from specific disorders if the nutrient supply falls below a minimum requirement for any particular constituent.

The supply of nutrients thus has a dominating effect on the performance and efficiency of agricultural animals: without an adequate supply there can be no production at all.

In general, the problem is one of supplying *sufficient* nutrients but rationing is nevertheless a common feature of animal feeding, either to limit the consumption of particularly expensive feeds or, in some cases, to limit total nutrient intake. The commonest examples of the latter are to prevent deposition of excessive fat (in pig production, for example, or in mid-pregnancy with sheep), either because it is unwanted in the product or because it may have a subsequent deleterious effect on the animal's health. Both represent inefficiencies, since they increase the inputs and add nothing to the desired outputs. Other reasons for wishing to limit feed intake include the need to restrict growth rate at particular times, in the interests of later, or whole life-time performance. This has been argued for in relation to the rearing of heifers, for example (Crichton *et al.*, 1959, 1960a,b), where overall performace may be increased if heifers are not grown too rapidly in their first year or so, although this depends upon age at first calving (Roy, 1969).

However, there are relatively few problems about restricting intake: in controlled feeding situations, the quantities fed can be closely determined and, under grazing conditions, the area available and the stocking density can both be manipulated to some extent. There may nevertheless be a difficulty in controlling the feed intake of individuals within groups and it is a common problem in animal production to determine whether any gains in efficiency are worth the additional trouble and cost of controlling feed intake on an individual basis. The more uniform the animals are within groups, the less need there is to feed differentially, but this may not solve problems due to inequalities of intake arising from bullying and other respects of animal behaviour.

The need to ensure the appropriate level of feed intake for individuals stems from the very great influence that this has on the efficiency of feed conversion. The latter is important biologically, because it determines the number of animals that can be sustained by a given food-producing area, and economically, because the cost of feed generally represents a high proportion of the total costs in animal production (Table 18.1).

TABLE 18.1   Feed costs as a percentage of total costs in animal production systems

Production system	% of total costs
Milk	49
Intensive beef[a]	29
Lowland suckler beef	46
Sheep	48
Pig breeding	77
Pig fattening	59
Eggs – hens	70

[a]Mainly bulls, 10–18 months old, fed on maize silage and concentrates, housed on slats

Sources: University of Reading farm's budgets for 1975–76 (unpublished data); MLC (1978); Nix (1976); Emmans and Charles (1977)

All this applies primarily to warm-blooded animals but the vast majority of agricultural animals are in this category. The reason for making a distinction between warm- and cold-blooded animals is the large difference between them in "maintenance" requirement, i.e. the energy required to maintain normal metabolic activities. In general, an animal fed at a maintenance level will neither gain nor lose weight, by definition, and this will be true of both warm- and cold-blooded animals. However, the maintenance requirements of the latter are much lower (see Table 18.2), especially at low temperatures. As the ambient temperature falls below the critical temperature, so the dietary energy needs of homeotherms increase, in order to maintain their characteristically high body temperatures (Table 18.3). The needs of poikilotherms, by contrast, fall with decreasing temperature and, at very low temperatures, may become very small indeed. Where food shortage is associated with low temperatures, therefore, poikilotherms tend to be relatively more efficient (or less inefficient) than homeotherms. The same may also be true where food shortage is associated with aridity, since many poikilotherms are able to "aestivate", i.e. enter an inactive state similar in many respects to hibernation in winter.

TABLE 18.2   The maintenance requirements of animals for energy

	Energy requirement per day (MJ per kg liveweight)
Warm-blooded	
Beef cattle	0.20
Fattening sheep	0.26
Pigs	0.19–0.22
Cold-blooded	
Trout	0.078

Feeds for cattle and sheep are assumed to have 10.9 MJ of ME per kg DM; 400 kg beef animals and 40 kg sheep are kept at temperatures above "critical".
The pig figures are for sows of 135 or 230 kg Lwt, in an environment compatible with good health.
The trout requirement is calculated from figures for brown and rainbow trout kept at 15°C.

Sources: A.R.C. (1965); A.R.C. (1966); Brown (1946); Huisman (1976)

The maintenance level of feed intake thus represents the quantity that has to be fed before any kind of production (growth, milk yield, egg output) is obtained. Thereafter, each increment of feed intake results in some production, up to a maximum value associated with a feed level termed maximum voluntary intake, with the result that the efficiency of feed conversion into product increases with increases in intake until the maximum is reached. On very concentrated diets, it is possible for this level of intake to result only in the deposition of fat and, if this is not wanted, the efficiency will decline at very high intake levels.

TABLE 18.3   Increased maintenance requirements of homeotherms at temperatures below their critical temperature

Animal	Body wt (kg)	Surface area (m²)	Critical temperature (°C)	Additional ME required per °C fall in temp. (kJ per day)
Beef cattle				
Baby beef				
1 kg gain per day	150	2.58	–12	557
	350	4.55	–12	873
1.3 kg gain per day	150	2.58	–15	570
	350	4.55	–26	873
Store cattle				
at maintenance	250	3.64	–16	630
0.4 kg gain per day	250	3.64	–30	670
Fat stock				
0.8 kg gain per day	450	5.39	–36	913
1.5 kg gain per day	450	5.39	–36	1 034
Beef cow				
at maintenance	450	5.39	–21	750
Dairy cattle				
dry, pregnant	500	5.79	–14	895
giving 2 gal per day	500	5.79	–24	960
giving 5 gal per day	500	5.79	–32	1 020
giving 8 gal per day	500	5.79	–40	1 066

Sources: Webster (1974); Webster *et al.* (1978)

The relationships underlying these response curves for agricultural homeotherms are illustrated in Fig. 18.1. In view of the practical importance of these relationships, it is not surprising that a great deal of work has been done in this area and a great deal of information is available.

Understandably, there is much less information relating to poikilotherms, but the situation can be illustrated with reference to fish (Fig. 18.2).

With homeotherms, most aspects of production efficiency are improved by increasing the level of performance, sometimes with an optimum value lying somewhat short of the maximum, unless this involves a change to inputs with a markedly higher unit cost.

Thus beef cattle may be grown more rapidly by changing from grass to a barley diet: this would improve the energetic efficiency of feed conversion but might reduce economic efficiency or even the efficiency of land use.

Similarly, performance of pigs might be improved by a better controlled environment resulting in improved feed conversion efficiency: but the cost of the additional environmental control might lead to a reduction in profit. There are times when it is cheaper to use part of the animal's feed to keep it warm. It is also a method for making use of (more or less) current solar radiation for heating, rather than depending upon fossil fuels.

This need to obtain high individual performance leads to two major difficulties, both of them most marked under grazing conditions. The first is caused by the fact that plant production is generally rather seasonal and the second by the problems of achieving an optimum stocking rate.

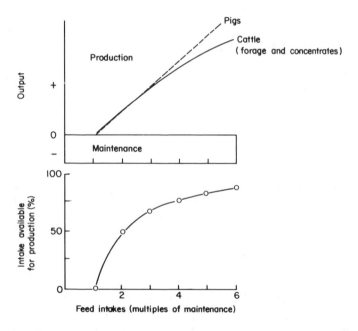

FIG. 18.1 An illustration of the relationships underlying response curves for homeotherms (from Balch and Reid, 1976)

Seasonality of Supply

This need not affect highly controlled animal production systems. Pigs and poultry that are fed on dry, stored feeds can be supplied with desired quantities, independent of the season. Fluctuations may nevertheless occur

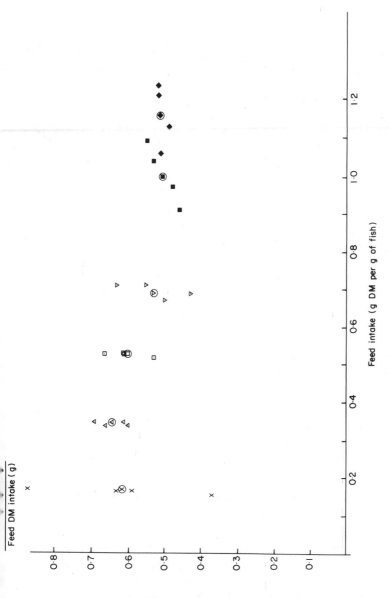

FIG. 18.2 A response curve for poikilotherms using Chinese grass carp as an example. Note: The data for this figure are from an experiment at University of Reading, using six treatments, each having four aquaria containing two fish, given different levels of feed. The points shown are for each aquarium plus a mean figure for each treatment, which is circled (J. Domaniewski, personal communication).

in the nature of the contributory components of the diet, due to variation in price or availability, but this should not necessarily affect the quantity fed or the level of performance obtained.

For grazing animals it is often quite different, mainly because the pattern of herbage growth is non-uniform. This is very marked in climatic regions that include very cold or very dry periods. In the UK, for example, a typical grass growth curve varies in rate from zero to over 80 kg DM per hectare per day (see Fig. 10.1, p. 123).

The feed requirement curves of agricultural animal populations are rarely anything like this. For groups of growing animals (e.g. a group of bullocks) the feed requirement increases linearly as the individuals get bigger: for dairy cows, it fluctuates with lactation and for sheep it varies chiefly with the reproductive cycle.

The result is that, even when the animal's annual feed requirement is adjusted in total to equal that of the annual feed supply, there are major discrepancies, giving rise to periods of surplus and shortage (Fig. 18.3).

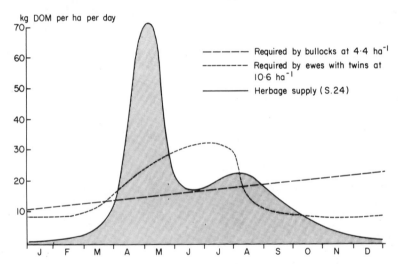

FIG. 18.3 Balanced total supply and demand (5800 kg digestible organic matter per hectare per year) (from Spedding, 1971)

If there is a practicable and economic means of transferring herbage from times of surplus to times of shortage, then the problem may largely disappear. This is not generally the case, however, and all methods of conservation involve some additional cost and some additional losses. The losses may reduce the quantity of nutrients preserved or may result in lowered quality in the conserved product.

Artificial drying of herbage is the best way of preserving quality and minimizing losses but it is by far the most costly, both in energy and monetary terms. Drying in the sun, to make hay, can be very effective but is very weather dependent and commonly incurs substantial losses of both quantity and quality. Ensilage can be quite simply achieved by sealing herbage within a plastic cover but losses are usually significant in most methods of ensilage and costs are not negligible.

One of these processes is generally employed, supported by supplementary feeding of cereals and protein concentrates (or by-products: e.g. see C.I.A.T., 1975) where required to raise the nutritional quality of the diet to what is needed. Another way of reducing discrepancies between supply and demand is by adjusting animal numbers (by selling or buying stock seasonally) or by adjusting stocking densities (by using a variable amount of land for grazing, according to the rate of herbage production). Conservation usually has some of these effects also, because areas put aside to grow conservation crops automatically reduce the area left for grazing: similarly, after conservation crops have been harvested, these areas are re-incorporated in the grazing cycle.

Conservation processes may be applied to a substantial proportion of the herbage grown, depending on the type of animal production system (see Table 18.4). The inefficiencies of the main conservation methods (Table 18.5), in the use of labour and energy and in the preservation of nutrients, therefore have effects on the total system that are related to the proportion of the total herbage production that is conserved.

TABLE 18.4   Proportion of herbage grown which is conserved

System	Herbage grown % grazed	% conserved	Source
Cattle (lowland single suckling)	40	60	A.D.A.S. (1970)
Sheep (fat lamb from leys)	62	38	Young and Newton (1975)

Quite apart from conservation inefficiencies, which may not be inevitable, the seasonality of herbage supply clearly exaggerates the difficulty of matching the quantity of feed available and the requirements of the animal population at any one time. It exaggerates but does not create the problem, since variation in the amount of herbage grown would occur both within and between years, as a result of variation in the weather (independent of seasonal effects). In grass growing situations, therefore, there is always a problem of choosing the "correct" stocking rate.

TABLE 18.5   Use of resources in herbage conservation

	Labour hours per tonne	Dry matter losses (%)(3) during conservation	Support energy(4) used MJ per 100 MJ feed gross energy
Hay			
field-cured	2.74(1)	30	18
barn-dried		15	37
Silage	0.69(1)	20	21
Dried grass	4.9–4.7(2)	5	136

Sources: (1) Milk Marketing Board (1973); (2) Theophilus (1979); (3) Wilkins (1975); (4) Walsingham (1976)

Optimum Stocking Rate

There are several extremely difficult problems associated with the choice of stocking rate. One is the fact that, even if the correct rate can be chosen, it cannot easily be altered to fit the changes in herbage supply from week to week. Although herbage does not have to be consumed on the day it is grown, there is a limited number of weeks before its quality will deteriorate (Fig. 18.4) and senescence and death occur within a relatively short time.

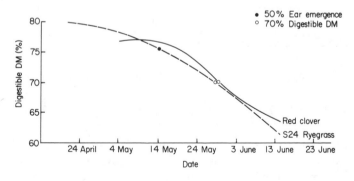

FIG. 18.4   The decline in herbage digestibility with age (after Spedding and Diekmahns, 1972)

There is a further major difficulty in ever knowing how much herbage is present: this presents a major sampling problem to the grassland scientist;

the practising farmer is poorly equipped to deal with it and usually relies upon a visual assessment. The latter does not, of course, commit him too far, since he can also judge when to remove animals from a paddock. He is therefore operating on a double judgement, of how many and for how long, and can adjust his judgement even within a day.

A more fundamental problem relates to the meaning to be attached to "optimum stocking rate". The concept has been discussed in detail by a number of workers (Mott, 1960; Spedding, 1971; Mott, 1973; Leaver, 1976), all of whom have recognized two major propositions. The first is that animal production per unit of land rises with increasing stocking rate (this must be so) up to a maximum value, beyond which production will decline (again, this must be so for warm-blooded animals). Even without any interactions between the presence of grazing animals and the growth rate of pasture, this proposition would be true for any quantity of feed supplied to homeotherms. If there are insufficient animals to consume all the feed, animal production will be less than the maximum possible. If there are too many animals, so that each one gets less than its maintenance requirement, not only will production be less than maximum, it may well be negative and the animals must lose weight and eventually die.

The second proposition is that, as the stocking rate increases the performance per individual animal will decrease. This is bound to be true above a given stocking rate, but the evidence is variable in relation to very low stocking rates.

These two relationships are illustrated in Fig. 18.5. In practice, they are modified by interactions between the grazing animals and the pasture (effects of treading, defoliation and fouling by excreta), between animals (behavioural influences of one animal upon another), between the grazing animals and small fauna (effects of biting flies on grazing behaviour; effects of parasites on health), and by the influence of management (methods of grazing management, shape of paddocks).

The use of the term stocking rate (number of animals per unit area) is satisfactory for changes of stocking on a given area. For between-area comparisons, however, it is too imprecise and "grazing pressure" (the relationship between animal numbers (or demand for feed) and the amount of herbage present (or available)) is more relevant.

However the relationship is expressed, the problem of selecting an optimum stocking rate is clearly difficult. Choice must depend upon whether output per head or per hectare is of primary importance, or whether some minimum value can be set on one or the other. This is really part of the common problem of deciding how best to define the product being produced.

Once defined, the efficiency with which it is produced per unit of land may involve a compromise between the efficiency with which herbage is grown, harvested and converted by the animal.

There will be a stocking rate at which the maximum amount of herbage is grown and, over the short term, this may differ from the stocking rate that maximizes the amount harvested. Over longer periods, however, it is possible to arrive at a relationship that allows a choice of stocking rate to maximize the production of harvested herbage per hectare. This may not be the same as the stocking rate at which each animal is fed at a level that allows it to convert its feed most efficiently (in the sense of maximizing output per unit of feed). The reason why it is often desirable to exceed the stocking rate at which individual efficiency is maximized, is because a greater quantity of herbage will then be harvested (and converted).

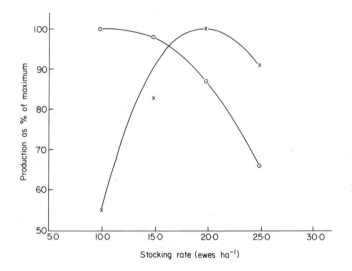

FIG. 18.5   An example of the relationship between production per hectare and growth rate of the individual animal (derived from Spedding, 1970). X = mean LW gain of lambs per hectare against stocking rate of ewes; O = mean LW gain per lamb per day against stocking rate of ewes (from Spedding, 1975)

This appears to be a good example of the impossibility of maximizing efficiency in all directions at the same time, although it is possible that a greater understanding of the underlying processes would allow this particular problem to be resolved.

In many cases, of course, there are several different animal requirements to be met, as in mixed grazing with sheep and cattle (Hamilton, 1976), mixtures of milking cows and rearing heifers, or ewes and their lambs. There are times when it is possible to distinguish clearly between the different needs of the animals grazing on the same area and it may then be practicable and economic to arrange for one group to take precedence over another, generally by a rotational grazing management that allows those animals with the higher requirement to graze a paddock first. High individual feed intakes can thus be obtained when they are most needed, yet a high proportion of the total herbage present is harvested, using the high grazing pressures that the following group can safely exert.

Throughout this discussion, a recurrent point has been the fact that agricultural animals cannot be regarded simply as individuals, or even groups of individuals. Breeding populations include animals in quite different physiological states, with markedly different needs and performance attributes. The proportions of male, female, old and young in such populations has a big influence on the efficiency with which resources are used, including feed, and for many aspects of efficiency the reproductive rate is the dominant factor.

References

A.D.A.S. (1970). Lowland single suckling. Profitable farm enterprises Booklet 4. M.A.F.F.

A.R.C. (1965). The nutrient requirements of farm livestock. No. 2. Ruminants. A.R.C. London. H.M.S.O.

A.R.C. (1966). The nutrient requirements of farm livestock. No. 3. Pigs. A.R.C. London H.M.S.O.

A.R.C. (1975). The nutrient requirements of farm livestock. No. 1. Poultry. 2nd edn, revised. A.R.C. London. H.M.S.O.

Balch, C. C. and Reid, J. T. (1976). The efficiency of conversion of feed energy and protein into animal products In "Food Consumption and Production", (Eds A. N. Duckham, J. G. W. Jones and E. H. Roberts), North Holland Publishing Co., Amsterdam.

Brown, M. E. (1946). The growth of brown trout, *Salmo trutta* (L.). III. The effect of temperature on the growth of two-year-old trout. *J. Exptl Biol.* **22**, 145–155.

C.I.A.T. (1975). Potential to increase beef production in tropical America. Proc. of a Seminar, Cali, Columbia, 1974. Series CE–N° 10.

Crichton, J. A., Aitken, J. N. and Boyne, A. W. (1959). The effect of plane of nutrition during rearing on growth, production, reproduction and health of dairy cattle. 1. Growth to 24 months. *Anim. Prod.* **1**(2), 145–162.

Crichton, J. A., Aitken, J. N. and Boyne, A. W. (1960a). The effect of plane of nutrition during rearing on growth, production, reproduction and health of dairy cattle. 2. Growth to maturity. *Anim. Prod.* **2**(1), 45–57.

Crichton, J. A., Aitken, J. N. and Boyne, A. W. (1960b). The effect of plane of nutrition during rearing on growth, production, reproduction and health of dairy cattle. 3. Milk production during the first three lactations. *Anim. Prod.* **2**(2), 159–168.

Domaniewski, J. (1980). University of Reading, Dept. of Agriculture and Horticulture. Personal communication.

Emmans, G. C. and Charles, D. R. (1977). Climatic environment and poultry feeding in practice. *In* "Nutrition and the Climatic Environment", (Eds W. Haresign, H. Swan and D. Lewis), Butterworths, London.

Hamilton, D. (1976). Performance of sheep and cattle grazed together in different ratios. *Aust. J. Expl Agric. Anim. Husb.* **16**, 5–12.

Huisman, E. A. (1976). Food conversion efficiencies at maintenance and production levels for carp, *Cyprinus carpio* (L.), and rainbow trout, *Salmo gairdneri* (Richardson). *Aquaculture* **9**, 259–273.

Leaver, J. D. (1976). Utilisation of grassland by dairy cows. *In* "Principles of Cattle Production", (Eds H. Swan and W. H. Broster), Butterworths, London.

Milk Marketing Board (1973). Fodder crops costs tables 1972–73. Economics Division.

M.L.C. (1978). Beef Improvement Services. Final results for suckler herds, 1977.

Mott, G. O. (1960). Grazing pressure and the measurement of pasture production. Proc. VIII Int. Grassld Congr. Reading, 606.

Mott, G. O. (1973). Evaluating forage production. Chap. 12 *In* "Forage Crops", 3rd edn, (Eds M. E. Heath, D. S. Metcalfe and R. F. Barnes), Iowa State University Press, Iowa.

N.A.S. (1963). "Nutrient Requirements of Beef Cattle", Revised edn, N.A.S. Washington, D.C.

N.A.S. (1968). "Nutrient Requirements of Dairy Cattle", 3rd revised edn, N.A.S. Washington, D.C.

Nix, J. (1976). "Farm Management Pocketbook", Wye College, Kent.

Roy, J. H. B. (1969). The nutrition of the dairy calf. *In* "Nutrition of Animals of Agricultural Importance", Part 2, (Ed. D. Cuthbertson), Pergamon, Oxford.

Spedding, C. R. W. (1970) "Sheep Production and Grazing Management", 2nd edn, Baillière, Tindall and Cassell, London.

Spedding, C. R. W. (1971). "Grassland Ecology", Oxford University Press, London.

Spedding, C. R. W. (1975). "The Biology of Agricultural Systems", Academic Press, London and New York.

Spedding, C. R. W. and Diekmahns, E. C. (1972). "Grasses and Legumes in British Agriculture", C.A.B. Farnham Royal, Buckinghamshire.

Theophilus, T. W. D. (1979). Personal communication, Agricultural Development and Advisory Service, Cambridge.

Walsingham, J. M. (1976). Energy for grass. Annual Report 1975 Grassland Research Institute, Hurley.

Webster, A. J. F. (1974). Heat loss from cattle with particular emphasis on the effects of cold. *In* "Heat Loss from Animals and Man", (Eds J. L. Monteith and L. E. Mount), Butterworths, London.

Webster, A. J. F., Gordon, J. G. and McGregor, R. (1978). The cold tolerance of beef and dairy type calves in the first weeks of life. *Anim. Prod.* **26**, 85–92.

Wilkins, R. J. (1975). Improving the nutritional efficiency of beef production. EEC Seminar, Theix, France.

Young, N. E. and Newton, J. E. (1975). "Grasslambs", G. R. I. Farmers Booklet No. 1. Grassland Research Institute, Hurley, Berkshire.

19

REPRODUCTION IN ANIMALS

In the previous chapter, the relationships between individual animal performance and efficiency of resource use (especially feed) were discussed. But it has been pointed out also that animal production is based on reproducing populations, rather than on collections of similar individuals, even if all phases within a population are not kept together or on the same farm.

The total resources used by such populations are not necessarily the same or in the same proportions as those used by the productive individuals but feed is usually a major cost. The efficiency with which feed is used is therefore at least as much concerned with populations as it is with individual animals and generally it is even more important.

The reason for the latter is that the rest of the population add greatly to the quantity of feed required (see Table 19.1) but little to the output of product.

The additional feed can be regarded as part of the maintenance requirement of the population and, as with individual animals, this has to be spread over as large a volume of production as possible, if feed conversion efficiency is to be high.

In the case of meat-producing animals, the volume of production depends primarily on the number of productive (e.g. growing) animals and the size that each finally attains. Size is determined by genetic considerations and by the age and rate of growth of the individual. The number of such individuals, however, is mainly a reflection of reproductive rate, modified by disease, accident and nutrition.

There are several ways in which reproductive rate can be expressed, the most important being the numbers of progeny (alive, total, at birth or later) per female (or per breeding female or per unit liveweight of ♀) or per head of population (± ♂) or per unit liveweight of the whole population, at one time, over one breeding cycle, over a year or over several years (or a life-time).

When comparing different species, it is necessary to take account of

TABLE 19.1 An example of the effect of feed requirements of the population on efficiency

Sheep	Input of feed, I (MJ)	Output of carcass, (MJ) O	$\dfrac{O}{I} \times 100$
Individual lamb, birth to slaughter[a]	2 143	225	10.5
Population[b]			
Ewe and 1 lamb	12 913	281	2.17
Ewe and 2 lambs	16 761	504	3.0
Ewe and 3 lambs	21 035	791	3.7

[a]Calculated taking an example of one lamb of twins with a birth wt. of 4.8 kg (energy content of 24.3 MJ), consuming 88 kg milk from the ewe (478 MJ) and 90 kg DM of herbage (1665 MJ) producing a caracse of 18.6 kg (249 MJ)
[b]Using data from experiment H683 (Grassland Research Institute) where ewes with singles, twins and triplets in indoor pens had feed intakes measured over two years

Sources: Young and Newton (1975); Penning *et al.* (1977); Gibb (1979)

differing $\male : \female$ ratios, breeding frequency, gestation lengths and longevity, all of which may vary substantially from one species to another (Table 19.2). There are often good reasons, also, for taking account of differences in body size, since the quantity of feed consumed by breeding animals is related to their size (Table 19.3); even so, larger mammalian species do not generally exhibit higher reproductive rates. Indeed, smaller species tend naturally to breed more frequently (but they also have shorter lives) and to give birth to larger litters (Table 19.4). This is an interesting phenomenon, since it is not obvious why larger species should not have large litters and, in fact, some do, such as the pig, with enormously beneficial effects on the agricultural usefulness of the animal and on its feed conversion efficiency. Nor is it clear why smaller species should have large litters, especially amongst the very small, since the birth weight of individual progeny then becomes very small indeed. Of course, some smaller species do have small litters.

The situation with birds is more variable, with no marked tendency for clutch size to be related to mature body weight (Table 19.5): egg size may also be related to body weight although in birds generally egg size varies from 2–27% of body weight (Lack, 1968).

Feed consumption by populations varies with the body weight of both males and females, and thus with their proportions, with the number of progeny and the size to which they grow (or are allowed to grow). Rate of

TABLE 19.2   Reproductive attributes of agricultural animals

Animal	Ratio ♂:♀ for breeding	Breeding frequency No. of times per yr or eggs per yr	No. of progeny per litter	Av. gestation or incubation length (days)	Longevity (yrs)
Cow	1:30–50	0.9	1	280	8–14
Cow (AI)	1: up to 234 000				
Sheep	1:30–40	1–1.5	1–3	147	5–8
Pig	1:20–25	1–2.2	8–12	114	2–3
Goat	1:40	1	1–3	150	6–10
Rabbit	1:10–20	7	8–12	31	2
Hen	1:12–20	180	—	21	2
Duck	1:5–8	110–175	—	28	3
Goose	1:2–5	25–50	—	30–33	5

Sources: Fraser (1971); Spedding and Hoxey (1975); Foote (1974); M.A.F.F. (1973)

TABLE 19.3   Liveweight and feed consumption of breeding animals

Animal	Wt of non-pregnant ♀ (kg)	Feed consumption[a] (kg DM per day)
Cow	500	4.2
Sow	230	2.3
Ewe	70	0.7
Rabbit	4.5	0.10
Hen	2.0	0.08

[a]for maintenance

Sources: A.R.C. (1965); A.R.C. (1966); Walsingham (1974); Morris (1974)

TABLE 19.4   Reproductive rate of animals of different sizes

Animal	Lwt of breeding ♀ (kg)	No. of pregnancies per yr	Av. size of litters
Elephant	3629	0.25	1
Horse	500–900	1	1
Cow	500	1	1
Sow	230	2	8–10
Ewe	70	1–1.5	1–3
Rabbit	4.5	7	8
Hamster	0.118	5[a]	10
Laboratory mice	0.025	7[b]	10

[a]In twelve-month lifespan
[b]In nine-month lifespan

Sources include: Petrides and Swank (1965); Lane-Petter and Pearson (1971); Worden and Lane-Petter (1957); Whitney (1963)

growth also influences the matter, as described earlier, and therefore all those factors that affect growth rate play a part.

The size of the parental species or breeds sets some limits to the upper size of the progeny but custom may prescribe an earlier stage and lighter weight for slaughter. Clearly, if animals are slaughtered too soon the quantity produced will be greatly reduced, while the parental feed intake remains the same, and the efficiency of feed conversion may be greatly lowered (Fig. 19.1).

If progeny are slaughtered at an optimum weight, from the point of view of efficiency of resource use, then one might expect size *per se* to have little effect, unless the effect of size on output is quite different from its effect on resource need.

Feed conversion efficiency in meat-producing suckler cattle, for example, may be little affected by cow size: the larger cow eats more but also produces progeny that grow to a larger size. Agriculturally, however, it is possible to cross males of a large breed with females of a smaller breed. This has the effect of increasing the final size of the progeny without increasing the feed intake of the cow (and the male is responsible for only a small proportion of parental feed intake).

In milk production, however, where larger breeds may produce more milk, the use of labour may be improved because the *numbers* of cows to be milked per unit of milk produced will be less. So the optimum size for an

TABLE 19.5 Body weight and clutch size in birds

Species	Mature body wt (kg)	Egg wt (g)	Clutch size (no. of eggs)	Clutch wt (g)	Clutch wt as % of body wt
Goose	6.5–10	135–215	9–10	1 015–2 150	15.6–21.5
Duck	2.3–4	60	10–13	600–780	19.5–26.1
Hen	2.5	58	13	754	30.2
Pheasant	0.9	30	10+	300+	33.3
Domesticated pigeon	0.4	19	2	38	9.5
Quail	0.1	8.8–10	10+	88–100+	88–100

Sources include: Gilbert (1974); Lack (1968)

agricultural animal may vary with species, climate, environment, function, nature of the product and the resource being considered.

For any particular size of meat-producing, warm-blooded animal it is likely that reproductive rate will influence efficiency, since it can so greatly affect output. Many of the inputs are either independent of reproductive rate or rise only slightly with it. Potential output, on the other hand, rises proportionately with reproductive rate, although actual output may be substantially less because of greater mortality, poorer nutrition (e.g. a shared milk supply by suckled progeny), poorer growth and smaller final size.

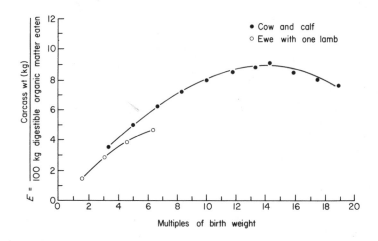

FIG. 19.1   Effect of product size (expressed as multiples of birth weight) on efficiency (E) (from Spedding, 1971).

In general, efficiency of feed conversion, for protein and energy, is therefore related to reproductive rate in a curvilinear manner (see Fig. 19.2), rising sharply from a reproductive rate of zero and gradually flattening off beyond some optimum value, itself dependent on a variety of circumstances.

One value of such curves is that they may indicate where the optimum does lie, for given circumstances, and whether higher reproductive rates are worth seeking or even breeding for. This can, of course, be extended, by theoretical calculations based on stated assumptions, beyond the data available for currently achievable levels. In extreme cases, this will only give rather negative information, of the form that, even given favourable assumptions (about mortality, growing rates, etc.), further significant improvement in efficiency is unlikely.

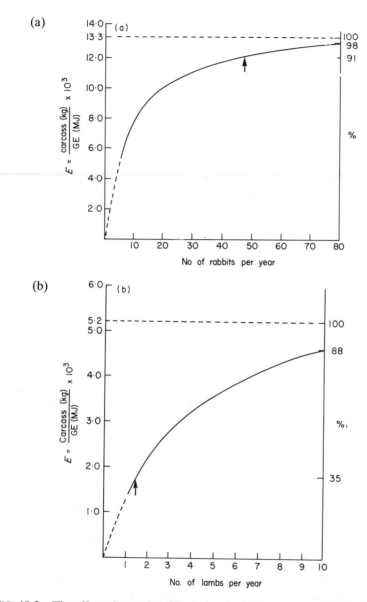

FIG. 19.2   The effect of reproductive rate on the efficiency of meat production in (a) rabbits and (b) sheep (from Large, 1976)

In Figs (a) and (b) the horizontal broken line at the top of the figures indicates the efficiency of an individual i.e. no allowances are made for the rest of the population. The point on the curve marked by the arrow indicates current levels of achievement. GE = gross energy; MJ = megajoules.

Methods of Reproduction

Reproductive mechanisms vary between species and they have considerable influence on the relationships between efficiency of resource use and reproductive rate.

The two most important mechanisms for agriculturally productive animals are mammalian parturition and egg production.

Mammalian parturition

Mammals give birth to live young after a gestation period that is remarkably constant for each species or, for agricultural animals, for each breed (see Table 19.6). This imposes a major limitation on the frequency with which birth can occur and acts as a constraint on reproductive rate. Gestation also means that there are physical limits to the burden of progeny and associated tissues and fluids that can be accommodated up to the point of birth. Table 19.7 illustrates this in terms of the percentage of a pregnant female's weight that is normally devoted to progeny. This upper limit on the total foetal burden influences the possible size of individual progeny, since this must then vary with litter size (see Fig. 19.3). Two major consequences flow from this. First, there is an upper limit to litter size because beyond some point each member of the litter becomes too small to survive. Secondly, the *total* output at birth is limited even below the maximum litter size.

These features of mammalian reproduction thus place limits on the potential reproductive rate that can be achieved and this is usually less than can be achieved by egg-producers (the exceptions are very small mammals, such as mice: the fact that it is not the live birth aspect that necessarily limits reproductive rate is illustrated by aphids, which can produce live young at over 120 per female (Baker and Turner, 1916)).

Egg production

Eggs are produced by all birds, most reptiles, amphibians and fishes and most invertebrates. Many of the last group are of major agricultural significance, but as pests, parasites and predators rather than as producers. This is not entirely true but few such animals are of world significance (the exceptions will be separately dealt with in later chapters).

Eggs are consumed as food directly, mainly from birds but also from fish and, for example, turtles, but the significance of egg production as a reproductive mechanism is much wider than this. For example, in warm-blooded animals used primarily for meat production, the constraints on reproductive rate are minimized, since eggs can be produced with great frequency (see Chapter 22). Thus the number of eggs per female per year can be extremely high and the maintenance overheads distributed over a large number of

TABLE 19.6   Gestation periods of mammals

Animal	Average gestation period (days)		Range[a]	Ref.
Cattle (*Bos taurus* and *Bos indicus*)	279		195–439	(1)
	280	Aberdeen Angus	273–282	
		Ayrshire	277–279	
		Jersey	278–280	
		Friesian	278–282	(2)
		Shorthorn	281–284	
		Hereford	283–286	
		Zebu	284–288	
Sheep (*Ovis aries*)	149		135–161	(1)
	147	Medium wool, mutton breeds	140–148	
		Long wool	146–149	(2)
		Fine wool	148–152	
Goat (*Capra hircus*)	150		124–168	(1)
	150		145–156	(2)
Horse (*Equus caballus*)	338		264–420	(1)
	336		325–341	(2)
Alpaca (*Lama pacos*)	240		—	(1)
	330		—	(1)
Buffalo (*Bubalus bubalis*)	314		287–336	(1)
Camel (*Camelus bactrianus*) (*Camelus dromedarius*)	396		333–432	(1)
Red deer (*Cervus elaphus*)	236		225–246	(1)
Llama (*Lama glama*)	170			
	180			(1)
	330			
Pig (*Sus scrofa*)	114		98–130	(1)
	114–115		110–117	(2)
Rabbit (*Oryctolagus cuniculus*)	31		27–37	(1)

[a]Some figures are for a single case and may be extreme

Sources: (1) Kenneth and Ritchie (1953); (2) Fraser (1971)

TABLE 19.7   Percentage of a pregnant female's weight that is normally devoted to progeny

Animal	Wt of non-pregnant ♀ (kg)		Wt. loss at parturition (kg)	%
Cow	590		72	12
Ewe	70	Single	8.5	12
		Twins	15.2	22
		Triplets	17.2	25
Sow	150		20	13
Rabbit	4	Litter of 10–12	0.6	16
		Litter of 5	0.3	9

Sources: Randall (1979); Newton (1979); Pauls and Whites (1978); Adams (1979)

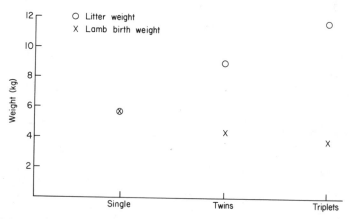

FIG. 19.3   Weights of individual progeny and of total litters in sheep (derived from Spedding, 1970)

products. This is so, of course, whether the eggs are consumed directly or not, but in the latter case they result in a very much larger total weight or volume of product.

Another feature of direct consumption is that the production of each unit of product takes such a short time that it does not add appreciably to the maintenance requirement. This is also true of milk production, where the product is removed daily, but both processes are quite different in this respect from growth.

For meat production, the unit produced is the carcase after slaughter (plus some other meat products) and this may take a substantial time to produce (weeks, months or years, depending on the species used, rate of growth and selected slaughter weight). During this time, each increment in weight, resulting from growth, adds to the maintenance requirement of the animal, so that the requirement in its last period before slaughter is vastly higher than just after birth (often by as much as ten times). Thus weight gained by a calf in its first few weeks adds to the daily feed requirement from that time until the animal is slaughtered, anything up to two years later.

This means that meat production by growth tends to be less efficient in the conversion of feed than milk production and sometimes than egg production, when these are at anything approaching normal levels of yield (see Fig. 19.4).

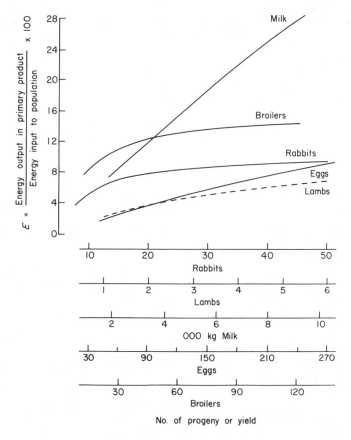

FIG. 19.4 Population efficiency of different species (from Spedding, 1975).

References

Adams, C. E. (1979). Personal communication. A.R.C. Institute of Animal Physiology, Cambridge.
A.R.C. (1965). The nutrient requirements of farm livestock. No. 2 Ruminants. A.R.C. London. H.M.S.O.
A.R.C. (1966). The nutrient requirements of farm livestock. No. 3 Pigs. A.R.C. London. H.M.S.O.
Baker, A. C. and Turner, W. F. (1916). Morphology and biology of the green apple aphis. *J. Agric. Res.* **5**, 955–993.
Foote, R. H. (1974). Artificial insemination. *In* "Reproduction in Farm Animals", 3rd edn, (Ed. E. S. E. Hafez), Lea and Febiger, Philadelphia.
Fraser, A. F. (1971). "Animal Reproduction Tabulated Data", Baillière Tindall, London.
Gibb, M. J. (1979). Personal communication. Grassland Research Institute, Hurley.
Gilbert, A. B. (1974). Poultry. *In* "Reproduction in Farm Animals", (Ed. E. S. E. Hafez), Lea and Febiger, Philadelphia.
Kenneth, J. H. and Ritchie, G. R. (1953). "Gestation Periods", 3rd edn, Technical Communication No. 5. C.A.B. of Animal Breeding and Genetics. C.A.B., Farnham Royal.
Lack, D. (1968). "Ecological Adaptations for Breeding in Birds", Methuen, London.
Lane-Petter, W. and Pearson, A. E. G. (1971). "The Laboratory Animal — Principles and Practice", A.P., London and New York.
Large, R. V. (1976). The influence of reproductive rate on the efficiency of meat production in animal populations. *In* "Meat Animals: Growth and Productivity", (Eds D. Lister, D. N. Rhodes, V. R. Fowler and M. F. Fuller), Plenum Press, New York and London.
M.A.F.F. (1973). Commercial Rabbit Production. Bull. 50. H.M.S.O., London.
Morris, T. R. (1974). Personal communication. University of Reading, Department of Agriculture and Horticulture.
Newton, J. E. (1979). Personal communication. Grassland Research Institute, Hurley.
Pauls and Whites (1978). The breeding unit. The management and feeding of breeding stock weaner pigs. Pauls publication No. 75. Pauls and Whites Foods Ltd., Ipswich.
Penning, P. D., Penning, I. M. and Treacher, T. T. (1977). The effect of temperature and method of feeding on the digestibility of two milk substitutes and on the performance of lambs. *J. Agric. Sci., Camb.* **88**, 579–589.
Petrides, G. A. and Swank, W. G. (1965). *Proc. 9th Int. Grassld Conf Brazil* **1**, 831–841.
Randall, E. M. (1979). Liveweight changes and condition scoring of dairy cows. M.A.F.F. Boxworth E.H.F. Annual Review, 1979.
Spedding, C. R. W. (1970). "Sheep Production and Grazing Management", 2nd edn, Baillière, Tindall and Cassell, London.
Spedding, C. R. W. (1971). "Grassland Ecology", Oxford University Press, Oxford.
Spedding, C. R. W. (1975). Energy and the environment: the supply of food. *In* "Energy and the Environment", (Ed. John Walker), Section 2 of Man and His Environment. A Symp. University of Birmingham 1975, University of Birmingham.

Spedding, C. R. W. and Hoxey, A. M. (1975). The potential for conventional meat animals. *In* "Meat", Proc. 21st Easter School in Agricultural Science, University of Nottingham, 1974. (Eds D. J. A. Cole and R. A. Lawrie), 483–506. Butterworths, London.

Walsingham, J. M. (1974). Personal communication. University of Reading, Department of Agriculture and Horticulture.

Whitney, R. (1963). Hamsters. *In* "Animals for Research", (Ed. W. Lane-Petter), Academic Press, London.

Worden, A. N. and Lane-Petter, W. (1957). The U.F.A.W. Handbook on the Care and Management of Laboratory Animals. 2nd edn, U.F.A.W., London.

Young, N. E. and Newton, J. E. (1975). Grasslambs. Farmer's Booklet No. 1, Grassland Research Institute, Hurley.

20

MEAT PRODUCTION

The main meat-producing species used in world agriculture are listed in Table 20.1. Numbers of animals only give an indication of importance, because animals vary enormously in size and productivity. Even so, the number of animals of any one species present in a geographical region (Table 20.2) is one indication of the scale on which the resources required by the species are being deployed.

TABLE 20.1   The main meat-producing species in world agriculture

Cattle
Buffalo
Sheep
Goats
Horses
Rabbits
Pigs
Hens
Ducks
Geese
Turkeys
Game animals and birds

Not all the products of meat-producing animals are meat and few animals produce only one product. Indeed, animal production has to be based on the purchase or production of breeding stock, to replace those that die, so live animals are always one of the products of a particular species, in addition to meat, offal, skins, hair or wool.

The main products are given in Table 20.3 with an indication of the

TABLE 20.2 Numbers of farm animals in different regions of the world, 1977 (1000 head)

	Cattle	Sheep	Goats	Pigs	Chickens	Ducks	Turkeys
World	1 212 861	1 027 859	410 343	666 274	6 335 085	155 167	89 832
Developed market economies	295 326	330 405	16 820	173 887	1 673 046	13 902	29 923
N. America	136 527	13 184	1 425	61 104	474 324	1 468	18 072
W. Europe	100 772	91 644	9 866	100 644	854 805	6 231	11 135
Oceania	41 017	194 150	42	2 757	49 656	365	431
Other developed economies	17 010	31 428	5 486	9 381	294 261	5 838	285
Developing market economies	702 770	433 257	319 351	114 755	2 022 521	63 137	22 026
Africa	130 618	109 169	111 641	7 082	429 700	5 727	1 494
Latin America	270 249	111 957	29 138	72 867	751 926	10 140	16 735
Near East	47 453	144 780	65 896	285	253 277	3 550	3 579
Far East	253 842	67 328	112 564	32 977	583 797	43 636	190
Other developing economies	608	23	114	1 546	3 822	85	29
Centrally-planned economies	214 766	264 197	74 171	377 632	2 639 517	78 128	37 883
Asian CPE	71 066	90 195	66 992	254 795	1 409 132	42 591	687
E. Europe and USSR	143 700	174 002	7 179	122 837	1 230 385	35 537	37 197

Source: F.A.O. (1978)

TABLE 20.3   The main products of meat animals (Information derived from the literature)

Animal	Carcase of progeny (kg)	Milk (litre yr^{-1})	Skin (m^2)	Fibre (kg yr^{-1})
Cattle	200–300	3 864–11 365	3.3–4.7	
Buffalo	144–279	900–3 182	3.3–4.7	
Sheep	18–24	150–1 200	0.6	1–9
Goat	4.3–8.4	600–2 045	0.4	0.23[a]
Horse	360	11–17 per day	3.3–4.7	
Camel	210–250	1 350–3 600	2.8	2.3–4.5
Rabbit	1–2		0.1	0.23[b]
Pig	45–67		0.9–1.1	
Hen	1.45–3.5			
Turkey	3–9			

[a]Cashmere goat
[b]Angora rabbit

quantities produced per animal in a "normal" situation (though in many cases this is a very variable condition and only a range can be given).

The major resources used in animal production are land, feed, fertilizer, labour, capital and water, though these are not independent of each other. It might seem unnecessary to consider, for example, both feed and land, since the one comes from the other. However, it may not do so entirely (e.g. fish meal) or it may be a by-product of another enterprise (e.g. barley straw). If it does come from land, it may not be the land occupied by the animals and, if it is at some distance, there is an additional cost of transport. Of course, even if the land is on the same farm, there may easily be an additional cost of resources used to harvest, process, store and transport the feed grown.

In grazing systems, such operations only apply to the conservation areas but in temperate countries, for example, the proportion of the herbage grown that is conserved may be quite high (see Table 18.4, p. 231).

In any event, the efficiency with which animals use their feed does not necessarily indicate efficiency of land use since this also depends upon the amount of feed grown per unit of land and this varies with the nature of the land. Thus a feed with a high feed conversion efficiency for pigs tells us nothing about the efficiency with which pigs could use land that cannot grow that particular feed. Furthermore, although land is often regarded as the ultimate limited resource, "land" is not a homogeneous commodity and some land is nearer to where the food is required, some land needs more inputs (of preparation, clearing, cultivation, fertilizer, labour, money or water) than other land, and all land could grow anything if the inputs were high enough and included sufficient control over the environment.

This argument simply reinforces the view that although the efficiency with which the major resources are used is important, no single efficiency ratio can be regarded as of over-riding importance, and judgement still has to be based on an examination of a whole range of them.

The orders of magnitude of the most important ratios are illustrated in Tables 20.4–20.12 for the main meat-producing species but, of course, each species can operate over a range of efficiencies, from zero up to some ceiling value (Spedding, 1966; Wilson, 1968). To obtain a picture of the ceiling values, what factors determine them and how closely they are achieved, it is first necessary to establish what are the essential characteristics of meat-producing systems.

TABLE 20.4 Efficiency values for output of carcase in relation to feed dry matter intake by meat-producing animals

	Feed intake of individual[a] (kg DM)	Carcase output (kg)	E^b	Sources[c]
Cattle	3 005	257	8.6	(1)
Sheep	108–136	19–18	17.6–13.2	(2)
Pigs				
(pork)	93.6	44.8	47.9	(3)
(bacon)	174.8	67.3	38.5	
Rabbits	2.37	0.99	42.0	(3)
Hens				
(broilers)	4.0	1.45	36.0	(3)

[a]Other than milk

$$^bE = \frac{\text{Carcase output (kg)}}{\text{Feed intake (kg DM)}} \times 100$$

[c]Sources: (1) Tayler (1970); (2) Treacher and Hoxey (1969); (3) Spedding and Hoxey (1975)

TABLE 20.5 Efficiency values for the output of energy in the carcase of independent meat-producing animals in relation to the input of feed energy (from Spedding and Hoxey, 1975)

	E^a
Beef cattle	5.2–7.8
Sheep	11.0–14.6
Rabbit	12.5–17.5
Pigs	
pork	35
bacon	35
Hens	16
Geese	13.4

$$^aE = \frac{\text{Total energy in the carcase}}{\substack{\text{Gross energy in the feed from} \\ \text{independence to slaughter}}} \times 100$$

TABLE 20.6 Efficiency values for populations of the animals listed in Table 20.5 (from Spedding and Hoxey, 1975).

	E^a
Suckler cows and calves	3.2
Sheep with	
singles	2.4
twins	3.4
triplets	4.2
Rabbits	8.0
Pigs	
pork	23
bacon	27
Hens	14.6
Geese	10

$$^aE = \frac{\text{Total energy in carcase produced}}{\substack{\text{Gross energy in feed for the progeny and a} \\ \text{proportion of the parents}}} \times 100$$

TABLE 20.7 Efficiency of conversion of dietary protein to muscle protein (adapted from Lodge, 1970, by Bowman, 1973)

	Muscle protein gain / Protein intake (%)
Cattle	
veal	13.3
beef	7.7
Sheep	
lamb	13.3
hogg	10.0
Pig	
pork	15.4
"heavy"	13.8
Chicken	16.6

TABLE 20.8 Efficiency of conversion of gross feed energy to edible protein and of feed protein intake to edible protein in lifetime performance of farm species (after Holmes, 1970)

	Edible protein / Feed protein intake (%)	Edible protein / Gross energy intake (g MJ^{-1})
Dairy herd	23	1.3
Dairy and beef herd	20	1.1
Beef herd	6	0.4
Sheep flock	3	0.2
Pig herd	12	1.0
Broiler flock	20	1.8
Egg-producing flock	18	1.9

TABLE 20.9   Relative efficiency (E) of energy production in grassland systems (derived from Spedding and Walsingham, 1975)

Production system	E^a for the use of			
	Land^e (ha)	Solar radiation^f (MJ)	"Support"^{g} energy (MJ)	Fertilizer N (kg)
Milk^b	15 576	0.000 47	0.54	129
Beef^c	5 772	0.000 17	0.11	22
Sheep meat^d	4 929	0.000 15	0.23	38

$^a E$ is expressed as energy output (MJ) of milk or carcases per unit of resource used, on an annual basis.
^{b}Permanent pasture dairy farm approximately 76 ha and 100 cows averaging 900 gal (4217 kg).
^{c}Suckler herd, intensive grassland system.
^{d}Lowland fat lamb system.
^{e}Assumed to be receiving nitrogenous fertilizer at the annual rate of (a) 121 kg ha^{-1} for milk, (b) 265 kg ha^{-1} for beef, and (c) 131 kg ha^{-1} for sheep meat.
^{f}Based on total annual radiation receipt of 33×10^6 MJ ha^{-1} yr^{-1}.
^{g}"Support" energy is defined here as the additional energy (labour, fuel and electricity) used on the farm, plus the "upstream" energy costs, i.e. those used to manufacture the major inputs (fertilizers, machinery, herbicides etc.; human labour is excluded) and the "downstream" energy costs of processing and distribution.

TABLE 20.10   An illustration of the relative costs of animal and crop production, expressed in terms of the major resources (from Spedding, 1976)

	Quantity of resource required to produce 1 MJ of gross energy in the product		
	Land (ha)	Solar radiation (MJ)	"Support" energy (MJ)
Wheat	0.000 017	563	0.19
Potatoes	0.000 01	328	0.2
Milk	0.000 114	3 762	1.8
Eggs	0.000 2	8 017	12.8

Calculated on an annual basis

Wheat, potatoes and milk calculations based on data in Spedding and Walsingham (1975)

Egg calculations were made using a maize/soyabean meal ration and "support" energy costs calculated by D. M. Bather, University of Reading

TABLE 20.11   The efficiency of land use for meat production (after Walsingham, 1972)

Species	No. of progeny per dam per year	$E = \dfrac{\text{kg carcase produced}}{\text{ha of land}}$
Cattle	1	598
Sheep	1	429
Poultry (broiler hens)	90	852
Rabbits	49	932
Pigs (pork)	18	812

Assumptions made:
Cereal yield = 3200 kg DM per ha per annum
Herbage yield = 9000 kg DM per ha per annum
Cattle and sheep consume herbage only
Pigs and poultry consume cereals only
Rabbits consume a diet containing 70% cereal and 30% herbage
For each species the food requirements of the dam and her progeny have been used in calculating the land requirement

TABLE 20.12   Efficiency of animal populations

Adult	Young per year	$E = \dfrac{\text{Gross energy in product}}{\text{Support energy input}}$
Suckler cows	1	0.11
Sheep	1	0.39
Rabbits	40	0.03
Hens	90–108	0.04

Calculated from Spedding and Hoxey (1975); Spedding, Walsingham and Bather (1976)

The Features of Meat Production Systems

The main features are shown in Fig. 20.1 for the entire meat-producing process; farm enterprises are often based on only a part of this. For simple-stomached animals, such as pigs and poultry, the growing, processing and utilization of feed are commonly quite separate businesses, conducted in different places and by different people. The majority of ruminants, on the other hand, are kept in systems of which feed production is an integral part. In both cases, however, it is quite common for different parts of the animal production process to be carried out in separate enterprises. The production of calves, for example, may be separated from rearing enterprises and these, in turn, may sell their products to "fattening" or "finishing" enterprises. The separation of rearing and "fattening" also occurs in sheep and pig production: poultry production for meat is usually separated from production of the eggs used as starting points and many animal enterprises depend upon the quite separate production of replacement breeding stock.

The major factors governing the efficiency of an enterprise will vary with the part of the whole system that has been separated off, but the essential determinants do not really change, only the way in which they are expressed. Thus an artificial rearing enterprise for calves will appear to be dominated by such factors as the feed conversion efficiency of the calves (determined largely by the nature of the feed and the growth rate of the calves), their initial cost and final value and the price of feed. This does not include anything about the efficiency of the breeding herd that produced the initial calves but all this is, of course, summarized in their cost. One of the effects of breaking a complete system up into constituent enterprises is generally to dilute the effect of biological processes that have gone before. The reason for this is quite simple: the price of the calf that represents the starting point for a rearing enterprise is not determined solely by the efficiency of the operations that produced it; it also reflects the supply and demand situation at the time of purchase. However efficiently a calf is produced, it will attract a low price if more are available than are required. Similarly, inefficient calf-production may still be rewarded, and impose a high starting cost on the rearing enterprise, if demand greatly exceeds supply.

The relevance of measures of biological efficiency may therefore also depend on the way in which a whole production system is sub-divided, since this determines the points at which external economic forces can impinge. This does not eliminate any of the factors that influence the system but it may modify (and not always reduce) the effect that they have.

The major factors determining biological efficiency can be derived from Fig. 20.1 and listed as follows:

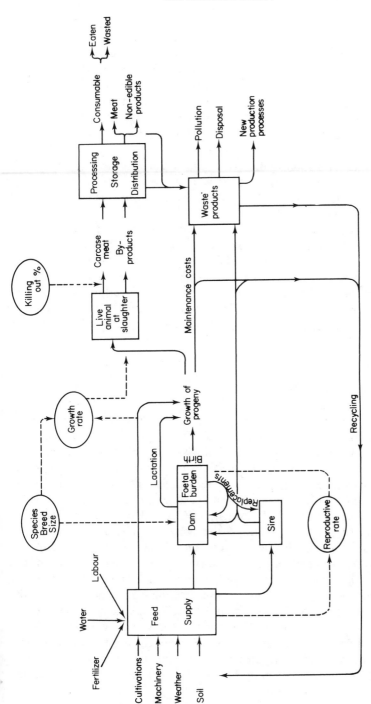

FIG. 20.1 The main features of meat production systems. Key: solid lines indicate flows of materials; broken lines indicate flows of information or the influence of the factors listed.

Species of animal
Size of dam and sire
Maintenance requirement
Sex ratio
Reproductive rate
Growth rate of progeny
Feed costs in gestation
Feed cost of lactation

Proportion of product(s) in the live
   animal at slaughter
Output of by-products
Outputs with a disposal cost
Labour requirement
Water requirement
Other special needs (including
   housing, veterinary attention, etc.)

Typical values for some of these attributes are given in Table 20.13 for the main meat-producing animals. (For a discussion on the energy cost of growth, see Millward *et al.* (1976); for the limitations to the manipulation of growth, see Elsley (1976).)

The effect of variation in these attributes on different efficiency ratios is illustrated in Figs 20.2–20.7. It is clear from the shape of the response curves that ceiling values can be predicted for many ratios and, in some cases, these values are already achievable. It is also clear that, although the effect of each factor can be examined separately and in sequence, it would be much easier to arrive at any outcome, including ceiling values, by using a more comphensive model designed to take account of all the main variables simultaneously.

Ceiling Values for Efficiency

One of the useful features of ceiling values is that they can be calculated on a range of theoretical assumptions in order to determine whether postulated improvements in animal production are worth seeking, in terms of the possible change in any aspect of efficiency that might be achieved. A number of potential ceilings have been calculated in this way (Table 20.14 uses the performance of individuals to indicate ceilings for energy efficiencies for populations and Table 20.15 uses most efficient known performances of protein efficiencies for populations) and many more could be worked out.

It is thus possible to distinguish between species, for example, and state that species A can never be as efficient (in some specified terms) as species B, since ceilings can be calculated as being much lower for the one than the other. Thus no amount of varying the determining factors will raise the efficiency of the one to the level of the other. Table 20.16 gives an example of this for sheep and rabbits. The changes that would have to be contemplated to alter things sufficiently would in effect produce a new species rather than change the attributes of the original one. However, changes in the management of an animal production system can be considerable and

TABLE 20.13 Typical values for some of the attributes of the major meat-producing animals

Animal	Mature size (kg liveweight) ♀	♂	Maintenance requirement per day of mature ♀ (MJ of ME[a] per head)	(kJ per kg Lwt)	Ratio ♂:♀ for breeding	Reproductive rate (progeny per ♀ per yr)	Growth rate of progeny (g per day) birth–slaughter	KO%[b]
Cattle	450–700	700–800	43.9–49.0	87.8–98[c]	1:30–50	0.9	600–800	55–60
Sheep	20–100	30–150	7.5–8.45	107–121[c]	1:40	1–3	200–350	45–50
Pigs	220	350	21–31	134–157[d]	1:20–25	20–23	380–480	70–75
Hens	3	4	0.8–1.1	405–679[d]	1:12	180	28	68–72
Rabbits	4	4	—		1:20	30–80	29	50–55

[a]ME = Metabolizable energy
[b]KO % = Killing-out percentage
[c]According to the quality of the feed, for 500 kg cow and a 70 kg sheep
[d]According to the size of the animal

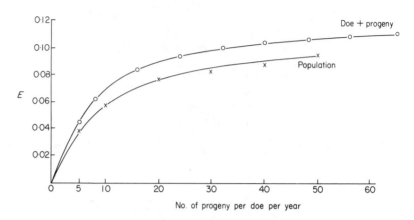

FIG. 20.2 Efficiency (E) for rabbits relative to reproductive rate

Does and Progeny

$$E = \frac{\text{Energy in carcass output}}{\text{Feed gross energy for doe and progeny}}$$

Population

In addition to the feed inputs to doe and progeny allowances were made for the buck, mortality rate and replacements, as follows: one buck to 20 does; mortality 0 for smallest litter, and up to 20% for largest; bucks replaced every three years; doe replacement, 50% per year.

The curves are hand drawn, not calculated.

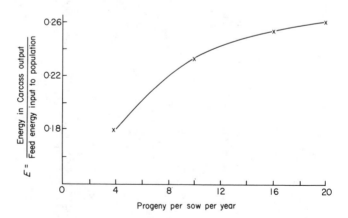

FIG. 20.3 Efficiency (E) for pigs relative to reproductive rate (from Spedding, 1981). Population data: one boar to 20 sows; sow replaced every two and a half years

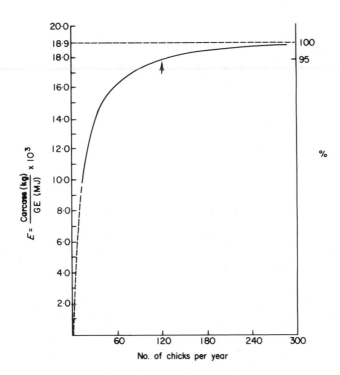

FIG. 20.4   The effect of reproductive rate on the efficiency of meat production in domestic fowl (from Large, 1976)

In Figs 20.4–20.6 the horizontal broken line at the top of the figures indicates the efficiency of an individual i.e. no allowances are made for the rest of the population. The point on the curve marked by the arrow indicates current levels of achievement. GE = gross energy; MJ = megajoules.

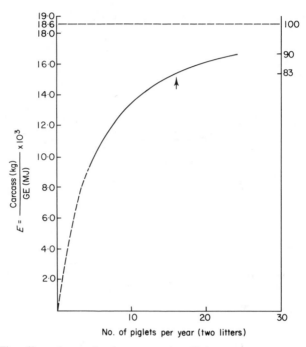

FIG 20.5   The effect of reproductive rate on the efficiency of meat production in pigs (from Large, 1976)

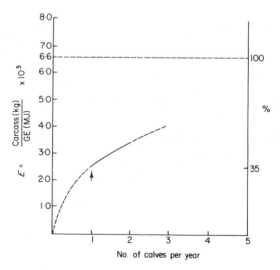

FIG. 20.6   The effect of reproductive rate on the efficiency of meat production in cattle (from Large, 1976)

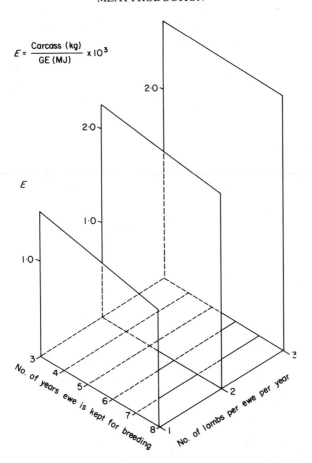

$$E = \frac{\text{Carcass (kg)}}{\text{GE (MJ)}} \times 10^3$$

FIG. 20.7   The effect of the replacement rate of breeding ewes on the efficiency of meat production in a flock (from Large, 1976)

can greatly influence efficiency. An example of this is the slaughtering of heifers after they have produced one calf (see Table 20.17).

All this, of course, presupposes that species can usefully be compared and only relates to conditions in which this is so. As has been stressed repeatedly, if a specific product, such as duck meat, is required, no other species can be more efficient at producing it. However, there are many important circumstances in which this is not so and in very few circumstances is it likely that a particular product (especially food) will be required regardless of the cost. Comparisons can thus be made where products are weighted according to their monetary values, even where a full economic assessment is not needed.

TABLE 20.14 Examples of E_F values for individual animals from independence[a] until slaughter (from Spedding, Walsingham and Large, 1976)

	Number of animals	Growth rate (g per head per day)	Liveweight at slaughter (kg)	$E_F = \dfrac{\text{Whole body gain (kg)}^b}{\text{Gross energy intake (MJ)}} \times 100$ Mean	SE of mean
Rabbits[c]	114	34	1.9	2.77	±0.16
Piglets[d]	22	533	93.0	2.43	±0.17
Calves[e]	72	1 064	385.7	1.26	±0.02
Lambs[c]	40	176	35.0	1.15	±0.07

[a]Weaning — except for pigs (one week after weaning at eight weeks of age)
[b]Liveweight has been used as the numerator largely because all the relevant energy contents were not available
[c]Data from the Grassland Research Institute
[d]Data from Dr R. Braude, National Institute for Research in Dairying
[e]Calculated from Owen, E. (1965) Intensive beef production: a study of the suitability of breeds and diets. Ph.D. Thesis, University College of North Wales, Bangor

TABLE 20.15 Conversion of annual feed protein to annual product protein at high productivity levels for different livestock classes

Class	Production level	Annual crude protein yield (kg)	Annual crude protein consumed (kg)	Efficiency (%)
Cow	6 800 litre at 3.25% protein	222	586	38
Hen	320 eggs at 13% protein	2.3	7	31
Broiler	6 crops at 1.8 kg liveweight	1.1	4	31
Fish	0.4 hectare of warm, carp pond	680	3 405	20
Rabbit	4 litters of 10 at 1.4 kg carcase	8	48	17
Porker	2¼ litters of 12 at 41 kg carcase	123	840	15
Lamb	2 litters of 3 at 16 kg carcase	11	125	9
Steer	295 kg carcase	34	568	6

From Wilson (1968)

TABLE 20.16 Theoretical efficiencies (E_F) for populations of 100 adult sheep and rabbits and their progeny (after Spedding, Walsingham and Large 1976).

$$E_F = \frac{\text{Gross energy in product}}{\text{Feed gross energy}} \times 100$$

Number of offspring per adult per year	Sheep	Rabbits
1	10.0	
2	11.2	
3	11.7	
4	12.0	
5	12.2	13.9
6	12.3	
7	12.4	14.9
9	12.5	
10	12.5	
16		15.7
24		16.1
32		16.4
40		16.5
48		16.6
56		16.6
64		16.7

TABLE 20.17 The feed conversion efficiency (E_F) of once-bred and not-bred female animals (from Spedding, Walsingham and Large, 1976).

	Not-bred females		Once-bred females and progeny		
	Age at slaughter	E_F	Age at slaughter Females	Progeny	E_F
Cattle	18 mths	4.5	30 mths	18 mths	3.6
Pigs	150 days	33.0	325 days	150 days	30.9

$$\frac{\text{Gross energy in carcase output}[a]}{\text{Feed gross energy}[b]} \times 100$$

[a]The calculations for once-bred females include one calf in the case of cattle and 8.2 baconers in the case of pigs
[b]Total feed for dams and progeny

This is illustrated in Table 20.18 for a comparison of protein production per unit of land by crops and animals. This is strictly more relevant to Part IV but makes the point more effectively than a comparison of one meat with another. However, in all such cases, it would soon be necessary to recognize that this was only a partial assessment, since the costs involved might be quite different. This would, of course, lead on to a full economic assessment and the simple point being made here is that, even in biological efficiency calculations, it is quite possible to take account of the fact that some products are more highly valued than others. Whether this is adequately reflected in current prices depends upon many other influences and it may be that it would be better, in this limited context, to recognize different

TABLE 20.18   Protein production per unit of land by crops and animals (after Spedding, 1980)

	Crude protein production[a] (kg ha^{-1})
Crop	
Grass (perennial ryegrass)	2 100
Cabbage	816
Field beans	
spring	678
winter	613
Peas	566
Potatoes (maincrop)	522
Wheat (winter)	469
Sugar beet	416
Maize	392
Rice	375
Barley	350
Animal	
Cattle	
beef (suckler)	53
beef (barley)	65
dairy — milk	118
Sheep	65
Pigs (baconer)	105
Rabbit	292
Hen	
meat	135
eggs	74

[a]For further details see Table 28.1

values by a weighting that involved no units, monetary or otherwise. Thus, if, for whatever reason, beef is more desirable or valued than poultry meat, it would be quite simple to say that this was so by a factor of two, or whatever value appeared right at the time. This would be one way of avoiding transient monetary values whilst recognizing that meats differ in water content, taste, flavour, nutritional value and palatability, in ways which no single expression can necessarily encompass.

The point is made here and will not be repeated for all other products, but it applies to them, of course, with equal force.

There are, however, some special features that only apply to meat production, to which attention needs to be drawn.

Special Features of Meat Production

Unlike milk and egg production, meat production always involves the slaughter of the animal. There is thus a clear end-point to the process: no product is harvested until this point and none can be produced afterwards. This point, in terms of animal age or weight or the time for which the process has been going on, may often be of great importance and is usually determined so as to optimize a number of efficiency ratios. Furthermore, this end-point is one of the features that is entirely under the manager's control, unless he runs out of feed or some other vital resource.

Another aspect of this same general property is that the product accumulates with time and, since it cannot be harvested, it has to be carried about by the animal that produced it, at some cost. The latter can be considerable. As the animal grows larger, its maintenance requirement increases vastly, yet, quite commonly, its rate of production may remain fairly constant. This is illustrated for beef production in Fig. 20.8, which shows the way in which the total maintenance requirement changes during the growth of animals in an 18-month beef production system and the way in which rate of production changes in the same period (and bearing in mind that these actual changes in growth rate are also the planned target rates chosen for the system).

Meat production is thus characterized by this increasing maintenance burden and, generally, this is not accompanied by proportionate increases in rate of growth. The result is that individual meat-producing animals tend to become less efficient in the use of feed as they grow larger. Clearly, there is no possibility, then, of slaughtering them before such efficiency declines. However, the efficiency of the whole family unit or population does not decline in this way, but increases as shown in Figs 20.2–20.7. Where enterprises are based solely on growing individuals this is not immediately clear, because the biological processes of reproduction and the feed costs associated with maintenance of the breeding stock are not included within the enterprise.

All this only applies, however, to animals in which the maintenance requirement is substantial, especially warm-blooded animals. It thus applies to both mammals and birds but to a much smaller extent to cold-blooded animals. Few of the latter have yet been used agriculturally for meat production, though this could change. Turtles are being exploited in an increasingly controlled fashion and there is some farming of prawns and shellfish (see Chapter 21). The major agricultural enterprises based on cold-blooded animals utilize fish, a topic of sufficient importance, and sufficiently different, to warrant a separate chapter.

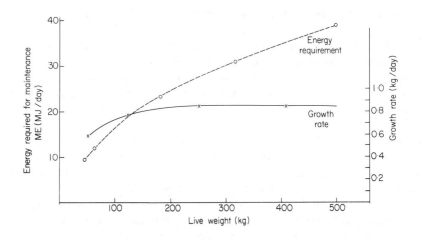

FIG. 20.8   Energy required for maintenance, and growth rate of beef cattle. Calculated from A.R.C. (1965) and J.B.P.C. (1971) for fasting metabolism.

References

A.R.C. (1965). The Nutrient Requirements of Farm Livestock. No. 2. Ruminants. Technical Reviews and Summaries. A.R.C. London.

Bowman, J. C. (1973). Possibilities for changing by genetic means the biological efficiency of protein production by whole animals. *In* "Biological Efficiency of Protein Production", (Ed. J. G. W. Jones), 173–182. Cambridge University Press, Cambridge.

Elsley, F. W. H. (1976). Limitations to the manipulation of growth. *Proc. Nutr. Soc.* **35**, 323–337.

F.A.O. (1978). Production Yearbook 1977, Vol. 31. F.A.O., Rome.

Holmes, W. (1970). Animals for food. *Proc. Nutr. Soc.* **29**, 237–244.

Joint Beef Production Committee (1971). Beef production, an intensive grassland system using autumn-born calves. Meat and Livestock Commission, Bletchley.

Large, R. V. (1976). The influence of reproductive rate on the efficiency of meat production in animal populations. *In* "Meat Animals: Growth and Productivity", (Eds D. Lister, D. N. Rhodes, V. R. Fowler and M. F. Fuller), N.A.T.O. Advanced Studies Institutes Series. Plenum Press, New York.

Lodge, G. A. (1970). Quantitative and qualitative control of proteins in meat animals. *In* "Proteins as Human Food", (Ed. R. A. Lawrie), 141–166. Butterworths, London.

Millward, D. J., Garlick, P. J. and Reeds, P. J. (1976). The energy cost of growth. *Proc. Nutr. Soc.* **35**, 339–349.

Spedding, C. R. W. (1966). Biological ceilings in sheep production. Paper presented to a Unilever Research Seminar on "The Scientific Implications of Sheep Intensification" at Colworth House, Bedford, Oct. 1966.

Spedding, C. R. W. (1976). Can human energy needs be met? *In* "Conservation of Resources", 83–95. Special Publ. No. 27, Chemical Society, London.

Spedding, C. R. W. (1981). New horizons in animal production. *In* "Animal Production Systems", (Ed. B. L. Nestel), Vol. 2A of World Animal Science. Elsevier, Amsterdam (in Press).

Spedding, C. R. W. and Hoxey, A. M. (1975). The potential for conventional meat animals. *In* "Meat", Proc. 21st Easter School in Agricultural Science, University of Nottingham, 1974 (Eds D. J. A. Cole, and R. A. Lawrie), 483–506. Butterworths, London.

Spedding, C. R. W., Walsingham, J. M. and Bather, D. M. (1976). Alternative animal species for more efficient farming. *In* "Energy Use and British Agriculture", (D. M. Bather, H. I. Day), Proc. 10th Annual Conf. of the Reading Univ. Agric. Club.

Spedding, C. R. W. and Walsingham, J. M. (1975). Energy use in agricultural systems. *Span* **18(1)**, 7–9.

Spedding, C. R. W., Walsingham, J. M. and Large, R. V. (1976). The effect of reproductive rate on the feed conversion efficiency of meat-producing animals. *Wld Rev. Anim. Prod.* **XXI(4)**, 43–49.

Tayler, J. C. (1970). Dried forages and beef production. *J. Br. Grassld Soc.* **25(2)**, 180–190.

Treacher, T. T. and Hoxey, A. M. (1969). Grassland Research Inst., Internal Report No. 170.

Walsingham, J. M. (1972). Ecological efficiency studies. 1. Meat production from rabbits. G.R.I. Technical Report 12.

Wilson, P. N. (1968). Biological ceilings and economic efficiencies for the production of animal protein, AD 2000. *Chem. Ind.* 899–902.

21

FISH PRODUCTION

Fish are an important component of the human diet in many parts of the world (Table 21.1) and a great deal of the fish caught and consumed by individuals never enters into trade and is presumably never recorded either.

Much of the fish consumed is caught at sea and this must be regarded as akin to hunting rather than as a part of agriculture. However, the catching of

TABLE 21.1 Examples of the fish component of the diet in different countries

| | Net food supply[a] *per capita* (g per day) | | | | | | | |
	Cereals and starchy foods	Sugars	Pulses, nuts, fruit and veg	Meat	Eggs	Fish	Milk	Fats and oils
Denmark 1970–71	422	137	323	170	30	48	731	79
Iceland 1964–66	251	150	121	191	9	107	754	67
Hong Kong 1964–66	368	60	327	111	27	58	52	32
Japan 1970	513	73	548	48	45	88	137	26
UK 1970–71	479	136	341	209	44	24	592	63
USA 1970	327	140	616	310	50	18	689	66

[a]Mainly in terms of fresh weight as purchased for household use

Source: F.A.O. (1972)

fish provides one of the most spectacular examples of the discrepancies that occur between different assessments of efficiency. Since fishing still exists as an important industry, it must be assumed that it can be efficient in financial terms. As a means of providing dietary energy or protein by the use of support energy, however, it appears to be grossly inefficient (Table 21.2 and Leach, 1976), although some calculations comparing farm animals and crops

TABLE 21.2(a)   Support energy requirements for UK cod fishing off the Faeroes

Year	Catch (tonnes)	kJ support energy used per kg fish landed	per g of fish protein	per kJ of fish energy
1924	40 823	5 157	30	1.6
1938	19 500	7 435	43	2.3
1946	25 594	2 943	17	0.9
1956	16 491	6 412	37	2.0
1961	1 689	10 785	62	3.3

Source: calculated from Slesser (1973)

TABLE 21.2(b)   The efficiency of support energy use in harvesting fish

Fish	Yield of dietary energy (MJ)[a] / Support energy used (MJ)	Yield of dietary protein (kg)[a] / Support energy used (MJ)
Sardines	3.38	0.08
Salmon		
pink	0.31	0.01
king	0.11	0.002
Cod	0.40	0.015
Flounder	0.38	0.013
Halibut	0.15	0.005
Haddock	0.20	0.005
Lobster	0.03	0.001
Tuna	0.08	0.0004
Shrimp	0.01	0.00053
Blue Crab	0.41	0.018

[a]Protein and energy values from Watt and Merrill (1963)

Calculated from Rawitscher and Mayer (1977)

with fish harvested from the sea (Rawitscher and Mayer, 1977; Krone, 1979) suggest otherwise. However, these have used the figures for feed energy plus support energy inputs to the animals and compared them with the support energy costs of harvesting fish from the sea with no feed energy costs. When calculated in this way the energy input per gram of protein output is much less for some forms of seafood production than for farm animal production.

Similarly, angling cannot easily be fitted into an agricultural framework, although it may be included (like caravan parks, holiday accommodation and the exhibition of rare breeds) amongst the enterprises on a farm. On the other hand, lakes and rivers may be stocked with fish in order to improve the fishing and this may involve the breeding of new stock and some feeding. There may therefore be some elements of the kind of control of the nutrition and reproduction of the animal population that are major characteristics of agriculture. As an enterprise, offering sporting facilities for sale, angling may contribute to the economic efficiency of a farm, and there is no reason why it should not also contribute significantly to food production, though it is hard to see any justification, from this point of view, for liberating fish to be caught again without further growth.

Fish farming implies a much greater degree of control over all aspects of production, but there are nevertheless many examples, especially in shellfish culture, where there remain considerable areas in which control is minimal. Both marine and freshwater fish are farmed but the degree of control is much less in the marine situation (Pullin, 1976; Purdom and Preston, 1977).

The main species of fish that have been utilized in fish farming are listed in Table 2.11, p. 26: by contrast with the main meat-producing mammals, a high proportion are carnivorous. In most cases, carnivores appear to be very inefficient users of resources for the production of human food and it is interesting to find that this does not seem to be so for fish. The main reason is that carnivorous fish are not necessarily eating other fish and, where they do so, such fish would not, in any case, be farmed. In other words, as with mammals and birds, once an animal is produced it can be consumed directly, just as readily as another, similar kind of animal, that might use it as feed. The case for producing an animal is either that we cannot utilize what it feeds on or that we would prefer not to do so. This is less likely to be the case where the two species are of the same kind (i.e. both mammals, or fish or birds) and produce the same kind of product (e.g. meat or fish). However, the inefficiency of a carnivore looks somewhat different if it is consuming animals that we could not easily consume (e.g. marine zooplankton) or would not wish to (e.g. earthworms or insects).

Since many of the carnivorous fish feed on very small organisms that, in turn, live on the very small plants that predominate in water, they are often the most suitable and attractive of the animal products available.

Furthermore, fish faeces tend to serve as a starting point for the growth of such organisms, so even herbivorous fish may usefully be kept with others of a carnivorous or omnivorous habit. Indeed, such integrated fish farming is likely to be the most productive in many situations.

Thus the aquatic situation differs in significant ways from the terrestrial one. Whatever the product, it will make more efficient use of resources if the food chain is short. This is well illustrated by the relative greater efficiency of herbivorous land animals than carnivorous ones and by the fact that more people can be fed per unit of land if they eat crop products directly. However, in aquatic conditions, the productivity of higher plants per unit area (and even more so per unit volume) of water may be very low. In fact, the biomass of plant material may often be lower than that of the primary consumers, largely due to the high rate of turnover of the latter at other times. Of course, the biomass of terrestrial animals, such as cattle and sheep, may be quite high in the winter, relative to the grasses on which they feed, but the contrast is never as extreme. Furthermore, to some extent the low biomass of grasses in winter is more apparent than real, since there is a substantial root mass below ground.

The very high reproductive rate of microscopic organisms has led to attempts to exploit them for the production of human food or for animal feeds (see Chapter 16) but their high rate of turnover demands frequent harvesting. This is possible in highly controlled ("industrialized") systems of production but would be impracticable and costly otherwise. In these circumstances, the conversion of micro-organisms by fish also has the effect of allowing an accumulation of nutrients, in the bodies of the fish, that can be harvested less frequently. Losses are relatively low, since waste materials are directly recycled and there is not the great heat loss associated with warm-blooded animals.

Use of Resources

Fish production uses resources somewhat differently from terrestrial systems. Fish must have feed, of course, but in fish farming this is not usually *grown* in the water. Crops *can* be grown in water and, with inputs of fertilizer, may produce large quantities per unit area: but fish production is not based on systems that do this in the same water, and usually involves feed grown on land.

Water is therefore primarily the medium in which the fish live (rather than *on* which they live) and in some ways more closely resembles the air than the soil of terrestrial systems. It is the source of oxygen but the available oxygen is at a much lower concentration than in air (Table 21.3) and the need to supply enough dissolved O_2 and to remove other gases leads to the use of water-circulating systems. The quantities of water used are vast and, if these are taken into account, production per unit of water is very low (Table 21.5) unless recirculation is used. This is also true of other animal production systems when the total water used is considered, including that transpired in the growth of feed crops (Table 21.4).

Feed conversion efficiency may be high in fish production (Table 21.6), especially in integrated systems using several species. This is true for both protein and energy, varying considerably, however, with the species used. Fish production can also be calculated in terms of land use (Table 21.7) for comparison with other forms of animal production. The higher yields shown depend, of course, upon substantial inputs of feed, with an additional land requirement.

TABLE 21.3   Available oxygen in air and water

Oxygen in the atmosphere	c.21%
Dissolved oxygen in water	c.2.8%

Source: Encyclopaedia Britannica (1959)

TABLE 21.4   Water requirements for agricultural production in arid zones

Unit of agricultural commodity	Total use of water per unit of production (litres)	Water supplied by irrigation per unit of production (litres)
1 lb (0.45 kg) wool	191 100	None
1 lb wool from winter pasture	28 210	15 925
1 lb wool plus 3½ lb (1.6 kg) meat	81 900	59 150
1 lb beef meat	19 110	14 560

Source: Hoare (1967)

TABLE 21.5   Production of fish per unit of water

Fish	System	Water requd per kg fish produced (litre)
Trout	Earth ponds (flowing water)	600 000–800 000
Trout	Swedish tanks (flowing water)	265 000–400 000
Common carp	Water-recirculating systems	
	one ton	240
	five ton	69
Carp, Tilapia, etc.	Pond — stagnant water (intensive culture)	5 000

Sources: Long (1972); Bardach *et al.* (1972); Hickling (1971)

TABLE 21.6   Feed conversion efficiency (F.C.E.) in fish production

Species	System of culture	F.C.E.
Common carp	Water recirculating system	
	one-ton	2.02:1
	five-ton	1.01:1
Catfish, *Clarias* spp.	Artificial propagation	5–6:1
Tilapia and *Hemichromis fasciatus*	Polyculture and artificial feeding in Togo	4–8:1
Tilapia and *Haplochromis mellandi*	Feeding plant feeds in Zambia	48:1
Trout	Commercial culture	
	in USA	1.5:1
	in Denmark	5–7:1
Pompano	Fertilized ponds USA	6.1:1
Black porgy	Cultured in Japan	3–4:1
Plaice or sole	UK (enclosed natural areas)	5:1

Source: Bardach *et al.* (1972) who make the following point regarding F. C. E.: "The fish culture literature contains more than a few reports of conversion efficiencies of 1:1 or only slightly more. Such reports, which appear to defy the second law of thermodynamics, are based on the dry weight of food and wet weight of fish. The best verified conversion efficiencies are on the order of 3–4:1, which is comparable to or slightly better than the best results obtained on land, where intensive feeding of animals is a much older process.

TABLE 21.7(a)   Approximate yield of various aquaculture systems

Species cultured	Location	Culture methods	Annual yield[a] (kg ha^{-1})
Channel catfish	South–Central United States	Commercial pond culture with heavy use of high-protein feeds	2 242–3 363 (best growers)
Channel catfish	Alabama	Experimental mono-culture using commercial catfish feed	1 412 (see next entry)
Channel catfish and *Tilapia mossambica*	Alabama	Experimental polyculture using same kind and amount of feed as in monoculture experiment (see entry above)	1 582, catfish; 224, *Tilapia* 1 806, total
Chinese carp (several species)	China and South-east Asia	Polyculture in heavily fertilized ponds, often integrated with terrestrial agriculture	Best ponds 7 062–8 071; average, 3 027–4 036
Common carp	Japan	Intensive pond culture with heavy feeding	5 045
Common carp	Indonesia	Growth in cages in streams polluted with sewage; no feeding	560 500– 840 750
Common carp	Japan	Culture in closed recircula-ting systems with heavy feeding	Up to 4×10^9
Milkfish	Taiwan	Intensive commercial culture with manuring; no feeding	1 906 (average) 3 027 (best growers)
Milkfish	Philippines	Intensive commercial culture with manuring or chemical fertilization; no feeding	1 009 (best growers)
Tilapia	Congo	Pond culture with virtually no management	up to 19 057 (mostly very small fish)
Wolffia	Thailand	Growing in unfertilized ponds	10 089
Yellowtail	Japan	Intensive commercial culture in floating cages in the sea; heavy use of high-protein foods	up to 62 776

[a]It must be noted that per-hectare yields for culture in streams, cages, and recirculating systems are computed on the basis of the area actually devoted to growing enclosures; real per-hectare yield would be considerably lower

Source: McLarney (1976)

TABLE 21.7(b) Approximate yield of various aquaculture systems

Species	Country	Technique/ site	Production (kg ha⁻¹)	Type of cultivation	Notes
Carp species	India	Pond	max. 110	Extensive, polyculture	
Carp species	India	Pond	300–900	Semi-intensive, polyculture	Feeding
Carp species	India	Pond	max. 2 800	Semi-intensive, polyculture	Stocking of seed only
Carp species	China	Lake	300	Extensive, polyculture	
Carp species	China	Pond	3 500–4 500 (max 15 000)	Semi-intensive, polyculture	Manure
Catfish (*Ictalurus* spp.)	USA	Cage (pond)	5 400	Intensive, monoculture	Feeding
Tilapia spp.	UAR	Lake	136–678	Semi-intensive, polyculture	Feeding, together with mullets and eel
Tilapia nilotica	Cameroon	Pond	980–3 225	Semi-intensive, monoculture	Feeding
Mullet (*Mugil cephalus*)	Taiwan	Pond	2 500–3 500	Semi-intensive, monoculture	Feeding
Milkfish (*Chanos chanos*)	Philippines	Pond	300–1 000	Semi-intensive, monoculture	Feeding, fertilizer
Milkfish	Taiwan	Pond	1 800–1 900 (max. 3 000)	Semi-intensive, monoculture	Feeding, fertilizer
Eel (*Anguilla japonica*)	Taiwan	Pond	3 500	Semi-intensive, polyculture	Feeding
Rainbow trout (*Salmo gairdneri*)	Taiwan	Pond	100–200	Extensive, monoculture	
Rainbow trout (*Salmo gairdneri*)	Taiwan	Pond	150 000–300 000	Intensive, monoculture	Feeding
Rainbow trout (*Salmo gairdneri*)	USA	Silo	max 6 500 000	Intensive, monoculture	Recirculating systems, feeding
Yellowtail (*Seriola quinqueradiata*)	Japan	Cage	max 280000	Intensive, monoculture	Feeding

Source: Ackefors and Rosen (1979)

The input of support energy varies with the system used and the climatic conditions in the environment in which it is carried out. Examples of the use of support energy and the efficiency with which it is used are given in Table 21.8. There is, of course, a possibility of using more solar energy to warm the water and to operate pumps but this has not advanced very far yet. In any case, fish farming tends to be very capital intensive (Shepherd, 1975).

TABLE 21.8   Support energy usage (per ha) in fish culture

	Support energy input $(GJ\ ha^{-1}\ yr^{-1})$	$\dfrac{\text{Protein yield (t)}}{\text{Support energy input (GJ)}} \times 100$
African *Tilapia* ponds	0.8	20
Philippine milkfish pens on lake	16.0	6
Japanese carp net cages	78.0	0.6
German carp ponds	15.0	0.5
British trout ponds	233.0	0.3
American catfish ponds	131.0	0.06

Source: Slesser *et al* (1977)

Growth and Reproduction

A great deal of attention has to be paid to hygiene in fish production and environmental conditions may be of great importance in fish breeding (the breeding of fish in terms of selecting better varieties is only just beginning, incidentally).

Nevertheless, reproductive rates are high (Table 21.9) and production systems are really concerned chiefly with growth. The difference between this and growing warm-blooded animals has much do do with the lower maintenance requirement of fish (Table 21.10) and results in less need to obtain high growth rates in order to obtain high feed conversion efficiency. Indeed, such efficiency is relatively little influenced by rate of growth: other features of resource use, such as equipment and capital, however, may require high growth rates.

All these features are similar for most forms of aquaculture but there are also important differences.

TABLE 21.9   Numbers of eggs produced by cultivated fish species

Fish	No. of eggs per ♀ per yr (thousands)	Sources
Carp, *Cyprinus carpio*	13.5–2 945	(1)[a]
Carp, *Cyprinus carpio*	180–530	(2)
Cod, *Gadus morhua*	2 000–9 000	(2)
Plaice, *Pleuronectes platessa*	16–350	(2)
Sole, *Solea solea*	150	(2)
Salmon, *Salmo salar*	10–100	(2)
Sturgeon, *Acipenser sturio*	800–2 400	(2)
Roach, *Rutilus rutilus* (L.)	4–11	(3)

[a]Sources: (1) Bardach *et al.* (1972); (2) Blaxter (1969); (3) Berrie (1972)

TABLE 21.10   Maintenance requirements of fish

Species	Wt (g)	Temp (°C)	Diet	Maintenance requirement (MJ per kg liveweight per day)
Brown trout, *Salmo trutta* (L.)	50	15	Meat and liver minced	0.078
Rainbow trout, *Salmo gairdneri Richardson*	67	15	Trout pellets	0.077
Carp, *Cyprinus carpio* (L.)	41	23	Trout pellets	0.072

Calculated from: Brown (1946); Huisman (1976)

The Culture of Crustaceans

Crustaceans are cultured (Table 21.11) on a considerable scale in some countries, the commonest example being the prawn, *Penaeus japonica*. (The terms shrimp and prawn are used here synonomously). Although this animal starts life in brackish water, it changes to fresh water before it becomes very large, so the largest part of the installation does not suffer from the corrosive effects of salt water. As with fish farming, the feed is produced elsewhere and fed to the prawns, which are kept in flowing water. Water use may therefore be substantial.

TABLE 21.11 Estimated world production through aquaculture of shrimps and prawns

Country	tonnes yr^{-1}
India	3 865
Indonesia	3 385
Japan	1 831
Malaysia	254
Philippines	2 543
Singapore	122
Thailand	2 543
Total	14 543

Source: Pillay (1973) quoted in Bell and Canterbery (1976)

Not a great deal of information is available concerning the efficiency of crustacean production but Table 21.12 illustrates the kind of results that have been obtained experimentally. Much poorer efficiencies may be found in practice. Bardach *et al.* (1972) note that the Kuruma shrimp culturist should expect to produce 1 kg of shrimp per 10–15 kg of feed (at the optimum temperature of 25°C). Poor feed conversion efficiencies are attributed to great losses of energy involved in moulting.

Shellfish Culture

The culture of shellfish is generally done in the sea or in estuarine waters, where feeding is based on filtering out organic matter derived from sewage outflows (Korringa, 1976; Bahr, 1976; Bayne, 1976).

TABLE 21.12 Growth performance of two typical *Macrobrac-
hium rosenbergii* adults

Age (days)	Food conversion ratio[a] ♂	♀
162	2.71	5.52
204	1.20	1.62
240	3.50	3.75
300	1.90	3.63

[a]Calculated as g dry food intake per g
body weight gain

Source: Walker (1975)

In these circumstances neither feed nor water use is of great consequence: the former would otherwise be wasted and has no value; the latter is entirely recycled.

The main resources employed are labour and capital, both very dependent on the precise nature of the operations and the methods used, and the area of water occupied, since this clearly incurs some cost.

Considerable labour may be involved in the initial establishment of the rafts, or beds used, and again in harvesting operations.

Yields per unit area can be enormous (Table 21.13) by comparison with terrestrial systems, even where the latter have all the feed delivered from elsewhere and the excreta removed. ("Yield" here applies to the meat but, in addition, there is an enormous yield of shell.) In some aquatic systems this

TABLE 21.13 Yields of shellfish per unit area

		Yield (kg per ha per yr)
Oysters	Japan (Inland sea)[a]	20 000
Oysters	U.S.A.	up to 5 000
Mussels	Spain[a]	300 000

[a]Raft-culture calculations based on an area 25% covered by rafts

Source: Bardach et al. (1972)

sort of difference is due to differences in the kind of organism used. In the case of shellfish, however, growth rates are not spectacular (Table 21.14) and feed conversion efficiency is not particularly high. The main features are the depth and volume of water-space that can be utilized without expensive supporting and housing structures, and the colossal flow of water and nutrients that are available.

TABLE 21.14   Growth rates of shellfish, (production probably includes the shell in all cases, but this is not always stated)

Species			Production	
Crassostrea angulata (Portuguese oyster)	France		65g incl. shell	in 3 years
C. gigas (Pacific oyster)	Japan (Inland Sea)		30–60 g	6–12 months
C. gigas	Japan (North)		30–60 g	18 months
Ostrea edulis (flat oyster)			65 g	4 years
Tapes japonica Asari clams			13 g	22 months
Anadara sp. "Cockles"			9 g	6–12 months
C. gigas	⎫		54.7 g	6 months
American hard clam	⎬ UK wt. increase between spring and autumn		2.69 g	6 months
Chilean mussel	⎭		6 g	6 months

Sources: Bardach *et al.* (1972); Walne (1974)

Since shellfish are relatively immobile, except in the very early stages, this method of farming is practicable: its chief danger (apart from storms) is from predators such as starfish and these present difficult control problems.

It is possible that much more use could be made of shellfish culture, perhaps in association with structures designed to harness wave or tidal movements.

References

Ackefors, H. and Rosen, C. (1979). Farming aquatic animals. *Ambio* **8(4)**, 132–143.
Bahr, L. M. (1976). Energetic aspects of the intertidal oyster reef community at Sapelo Island, Georgia (U.S.A.). *Ecology* **57(1)**, 121–131.

Bardach, J. E., Ryther, J. H. and McLarney, W. O. (1972). Aquaculture. Wiley-Interscience, New York.

Bayne, B. L. (1976). "Marine Mussels: Their Ecology and Physiology", (Ed. B. L. Bayne), Cambridge University Press, Cambridge.

Bell, F. W. and Canterbery, E. R. (1976). "Aquaculture for the Developing Countries", Ballinger, Cambridge, MA.

Berrie, A. D. (1972). "Productivity of the River Thames at Reading", *Symp. Zool. Soc. Lond.* No. 29, 69–86. Academic Press, London and New York.

Blaxter, J. H. S. (1969). Development: eggs and larvae. *In* "Fish Physiology" Vol. III (Eds W. S. Hoar and D. J. Randall), 177–252. Academic Press, New York and London.

Brown, M. E. (1946). The growth of brown trout, *Salmo trutta* (L.). III. The effect of temperature on the growth of two-year-old trout. *J. Exptl Biol.* **22**, 145–155.

Encyclopaedia Britannica (1959). Encyclopaedia Britannica Limited, London, Chicago, Toronto.

F.A.O. (1972). Production Yearbook 1971 Vol. 25. F.A.O., Rome.

Hickling, C. F. (1971). "Fish Culture", 2nd edn, Faber and Faber, London.

Hoare, E. R. (1967). Irrigation in semi-arid regions. *Outlook on Agriculture* **5(4)**, 139–143.

Huisman, E. A. (1976). Food conversion efficiencies at maintenance and production levels for carp, *Cyprinus carpio* (L.), and rainbow trout *Salmo gairdneri* Richardson. *Aquaculture* **9**, 259–273.

Korringa, P. (1976). "Farming the flat oysters of the genus *Ostrea*", Vol. 3. "Developments in Aquaculture and Fisheries Science", Elsevier, Amsterdam.

Krone, W. (1979). Fish as food. *Food Policy* **4(4)**, 259–268.

Leach, G. (1976). "Energy and Food Production", IPC Science and Technology Press, London.

Long, D. (1972). First catch your water. *Farmers Weekly* February 18th (IX).

McLarney, W. O. (1976). Aquaculture: toward an ecological approach. *In* "Radical Agriculture", (Ed. R. Merrill), New York University Press, New York.

Pillay, T. V. R. (1973). The role of aquaculture in fishery development and management, paper presented before the Technical Conference on Fishery Management and Development, Vancouver, 13–23 February 1973.

Pullin, R. (1976). Has U.K. marine fish farming lost its way? *New Scientist* **72**, 704–705.

Purdom, C. E. and Preston, A. (1977). A fishy business. *Nature (London)* **266**, 396–397.

Rawitscher, M. and Mayer, J. (1977). Nutritional outputs and energy inputs in seafoods. *Science (N.Y.)* **198**, 261–264.

Shepherd, C. J. (1975). The economics of aquaculture — a review. *Oceanogr. Mar. Biol. Ann. Rev.* **13**, 413–420.

Slesser, M. (1973). Science and the cod war. *New Scientist* **57**, p. 44.

Slesser, M., Lewis, C. and Edwardson, W. (1977). Energy systems analysis for food policy. *Food Policy* **2(2)**, 123–129.

Walker, A. (1975). Crustacean aquaculture. *Proc. Nutr. Soc.* **34**, 65–73.

Walne, P. R. (1974). "Culture of Bivalve Molluscs", Fishing News (Books) Ltd, West Byfleet, England.

Watt, B. K. and Merrill, A. L. (1963). "Composition of Food", Handbook No. 8, U.S.D.A.

22

EGG PRODUCTION

A wide range of animals produce eggs, but agricultural exploitation has been virtually confined to birds. Many other eggs are eaten, including those of insects, turtles and fish, and some of them are regarded as being great delicacies, but they are in the main collected. This is also true of a large number of wild bird eggs, and the gathering of eggs from populous sea-bird colonies has often been an important means of supplementing the diet of fishing people. This illustrates an aspect of efficiency that involves Man directly.

The biological efficiency with which sea-birds produce eggs from fish, judged as a production process using resources of sea-water (area or volume) and the fish-feed grown in it, is very low (Table 22.1). This is what

TABLE 22.1   Biological efficiency of sea-birds producing eggs from fish

Energy content of eggs per hectare of sea	$\dfrac{56\ 125}{470\ 000}$	$= 0.119\ \text{MJ}$
Energy content of eggs (MJ) per MJ of primary production in 4 700 km^2	$\dfrac{56\ 125}{19\ 664\ 800\ 000}$	$= 0.00000285$
Energy content of eggs (MJ)/ 28.6% of primary production	$\dfrac{56\ 125}{5\ 624\ 132\ 800}$	$= 0.00000997$

Calculated from Furness (1978) for sea-bird communities on Foula (an island) and the area of sea (4700 km^2 or 470 000 ha) utilized in foraging.

It is assumed that half of the breeding individuals are female, that the birds utilize 28.6% of the fish produced in this area and that the primary production is 1000 kcal m^{-2} yr^{-1} (4184 kJ).

No account is taken of juvenile birds or foraging outside this area in the non-breeding season.

Other sources: Camp *et al.* (1974); Cott (1954); Lack (1968); Richlefs (1977)

one would expect of carnivores, especially when they are often *secondary* carnivores, living on fish that are themselves carnivorous. However, the alternative of direct fishing by men in boats may be highly dangerous, costly and inefficient, especially in the use of support energy (Table 21.2). Egg production by sea-birds uses no support energy and very little may be needed in the collection process. The latter may also be hazardous, particularly from islands with steep cliffs and rough weather, but less so than fishing. This situation actually existed in the island of St Kilda, off the west coast of Scotland (Steel, 1965).

TABLE 22.2 Composition of edible portion of eggs of domesticated poultry

| | Average wt (g) | % of fresh wt | | | Energy content per 100 g (kJ) |
		Water	Protein	Fat	
Duck	67	70.1	13.5	14.5	774
Fowl	52	73.7	12.9	10.9	619
Goose	177	70.0	13.9	13.7	766
Turkey	72	73.7	13.3	11.4	649

Source: Bolton and Blair (1974)

Only birds have been used on any scale for the deliberate production of eggs for food, although this might change in the future, and only a rather narrow range of birds have been so used. The reasons may have to do chiefly with the eggs or with the habits of the birds themselves.

Bird eggs have a number of remarkable properties. They are already packaged in a very convenient and hygienic form, and can be stored for substantial periods without special conditions. Their shells are remarkably strong for their weight and yet can be readily broken open. They are transportable, with some care, and can easily be cooked. Apart from the shell, they are entirely edible, very digestible and with a high nutritional value (Table 22.2). Eggs tend to be deposited in particular places and birds can readily be trained or encouraged to use nest boxes. They are produced at short intervals, often over prolonged periods, and are easily harvested.

The species of birds that have been used may have been selected for their size, directly, since the birds are also eaten, and indirectly because of the convenience value of an appreciably-sized egg. Birds would also have to be selected that did not require one male for every female (although pigeons are of this kind: perhaps this is why they were chiefly used for meat production) and, indeed, did not require the presence of a male at all in order to stimulate egg-laying. These features are major variables amongst bird species and, for egg production, the necessity of a male presence is a major source of inefficiency, since it increases the need for all resources without increasing the number of eggs. The majority of agricultural birds require only a low proportion of males, even for breeding purposes (see Table 22.3 for the major attributes of the most important agricultural birds).

TABLE 22.3   Major attributes of the most important agricultural birds

	Mature liveweight (kg) ♂	♀	Age at sexual maturity (months)	Ratio for breeding ♂ : ♀	No. of eggs per yr	Carcase wt (kg)
Chicken for						
eggs	3	2	5–6	1:20	230–270	
meat	4	3		1:12	180	1.45–3.5
Turkey for						
meat	13–23	8–12	7–8	1:12	40–100	3–9
Duck for						
eggs	2.0	1.8	6–7	1:5–8	330	
meat	4.5	4.0		1:5–6	110–175	2–3
Goose for						
eggs		4.5	9–10	1:5	60	
meat	5–10	8–12	10–12	1:2–3	25–50	4–5

It may seem surprising that some of the most successful birds have not been used agriculturally. The house sparrow (*Passer domesticus*), for example, is successful in a very wide range of environments, lives on a variety of cheap feeds (including much household waste), clearly welcomes a degree of contact with human habitation that would make domestication easy, and it breeds readily. The enormous numbers of this species, spread all over the world, suggest that farming it would have been highly successful. Much the same may be said of several other species, such as the European starling (*Sturnus vulgaris*).

Similar arguments could be advanced about birds that are major agricultural pests, such as the red-billed quelea (*Quelea quelea*), which survive and multiply in spite of considerable efforts to reduce their numbers or even to eradicate them. The numbers of such birds in the wild are enormous: there are estimated to be between 10^9 and 10^{11} quelea in Africa, for example (Murton and Westwood, 1977).

The fact that they may live on foods we could eat ourselves cannot be the reason, since that also applies to the majority of the agriculturally-used birds. Indeed, originally, this may have been a major reason *for* domestication, so that the birds could be used as scavengers amongst the crop residues and around the household.

It may be that these birds are (or were) too small to be useful. Small size does have disadvantages, in terms of how much meat can be obtained from one animal, the handling of large numbers of eggs per unit of weight produced, the greater difficulties and cost of caging and, perhaps, a high metabolic rate.

This last point emerges very strongly from biological studies on animal size but, in fact, there is no obvious relationship between efficiency of production and size (Fig. 22.1). It is generally true that small animals have a

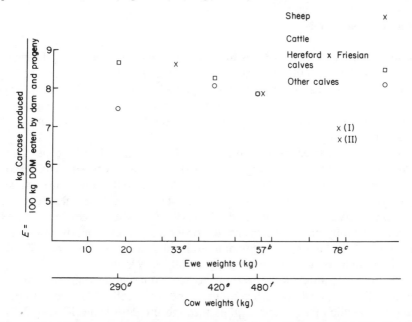

FIG. 22.1 Efficiency of production and dam size (after Baker *et al.*, 1973). Sheep: [a]Welsh Mountain ewe × Suffolk ram; [b]Kerry Hill ewe × Suffolk ram; [c]I Border Leicester × Cheviot ewe × Suffolk ram; II Devon Longwool ewe × Suffolk ram. Cattle: [a]Jersey cow and Hereford × Friesian calf; Jersey cow and Angus or Charolais × Jersey calf; [e]Welsh Black cow and Hereford × Friesian calf; Welsh Black cow and Hereford or Charolais × Welsh Black calf; [f]South Devon cow and Hereford × Friesian calf; South Devon cow and Hereford or Charolais × South Devon calf.

high surface volume/weight ratio and therefore tend to lose heat rapidly. But this depends upon ambient temperatures (and matters little in very hot climates), behaviour (such as nesting, altering body shape by curling up, huddling together), on fur and feather covering and many other factors.

Furthermore, although feed intake tends to be high, efficiency of production may not be lowered, if the rate of performance (growth or reproduction) is also high. Relative growth rates do tend to be very high in small animals and reproductive rates are also often high. This is particularly so amongst birds, where the restrictions of pregnancy and a fixed gestation period do not apply.

It may also seem surprising that so few herbivorous birds have been used or that no means has been found of exploiting the vast numbers of wading birds that feed in estuarine waters. The problems of containment should not be insuperable; quite simple horizontal wires will prevent flight by birds that require a long, gradual flight path, for example.

Herbivorous Birds

None of the herbivorous birds have developed very specialized digestive systems to deal with fibrous diets, and their digestive efficiency appears to be lower than that of adapted mammals, especially ruminants. There may have been many reasons for this but the most obvious is that such adaptation usually involves a considerable increase in weight and volume, which would be inimical to flight.

Ruminants, for example, have very bulky additional "stomachs" and, when these are full of liquid digesta, the extra weight may be equal to 20% of their empty body weight. Of course, this is also related to the ability to take in a large quantity of feed and then to digest it later, and different patterns of feeding behaviour can be imagined. Nevertheless, where digestion involves fermentation by micro-organisms, the process requires time, during which the mass of feed has to be accommodated and, since the digestion of fibre is rarely of high efficiency, the mass is generally very large.

Herbivorous animals that do not possess voluminous alimentary tracts tend to live primarily on cell contents rather than cell walls, and eat large quantities which pass rather rapidly through the gut (Table 17.5, p. 212). This is true of herbivorous birds, such as geese, and of herbivorous fish, such as the Grass carp (*Ctenopharyngodon idella*). The relatively inefficient digestion process may not matter too much if the faeces are recycled in some fashion, but the amount of herbage processed does have to be large in relation to the amount of product produced.

The feed conversion efficiency of geese for egg production is therefore low (Table 22.4), but this is partly due to the low egg output. That this is not a complete explanation can be seen from Fig. 22.2, in which egg number per bird has been extrapolated beyond anything yet achievable. It is probable that a ceiling value exists for efficiency of egg production by the breeds and varieties of geese used in the calculation, though this depends upon whether

TABLE 22.4   Feed conversion efficiency of geese for egg production

No. of eggs per yr	Feed intake (kg DM per bird per yr)	kg feed DM consumed per egg
20	82	4.1
40	85	2.1
60	89	1.5

Note: Very little data available for feed intake of geese. These calculations are theoretical but based on information available.

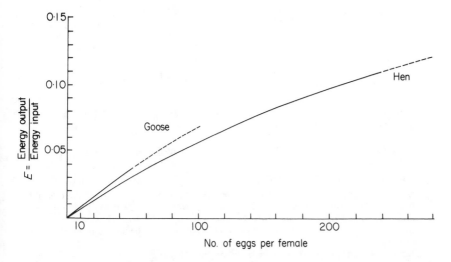

FIG. 22.2   Efficiency of energy production as eggs by the hen and the goose, related to the number of eggs produced per female per annum. Solid lines refer to levels of egg production per annum that are readily achieved on a flock basis. Efficiency is calculated as the energy produced per unit of feed energy consumed per annum by one female bird, and a proportion of the feed required to support one male.

The female/male ratios used were: 12:1 and 4:1 by hens and geese respectively (after Spedding, 1976).

energy or protein output is considered and cannot exclude the possibility that greatly improved varieties could be selected. Similarly, no serious consideration has yet been given to the deliberate selection of birds for efficiency at digesting herbage. Such birds might well be rather large and bulky and not very good at flying, but these are not necessarily disadvantages from an agricultural point of view.

The Use of Resources

The efficiency with which land is used for egg production by birds is high for grain-feeding birds (on land that can produce grain), relative to that of cattle fed grain but not generally higher than that of pigs on comparable diets from weaning to slaughter (Table 22.5). On land that can only grow herbage,

TABLE 22.5   Efficiency of land use by hens, cattle and pigs on mainly cereal diets

	DM intake (kg)	Number per hectare	Production per animal	Production per ha
Hen				
USA		85	240 eggs	1 183 kg
	40–45			20 400 eggs
Europe		50	240 eggs	696 kg
				1 200 eggs
Beef cattle	1 710	2	230 kg carcase	460 kg
Pigs				
pork	94	35	45 kg carcase	1 575 kg
bacon	175	18.5	67 kg carcase	1 240 kg
Broiler chicken	4	810	1.45 kg carcase	1 174 kg

Assumption made for the calculation:

Laying hen   Assuming ration of 72% maize; 18% soyabean meal; 10% minerals and vitamins. Assuming yields of maize, 4940 kg ha^{-1} USA; 2920 kg ha^{-1} Europe. Assuming yields of soyabean, 1810 kg ha^{-1} USA; 930 kg ha^{-1} Europe. Assuming intake of 40–45 kg DM per yr. Conclusion: USA, 85 hens per hectare. Europe, 50 hens per hectare.
Output: 240 eggs per bird per yr, each egg weighing 58 g.

Beef cattle   Intensive cereal beef system; DM intake from birth to slaughter estimated at 1710 kg. If this was all barley with a yield of 3239 kg ha^{-1} then it would be possible to rear to slaughter wt almost 2 (1.89) beef animals per yr each having a carcase wt of 220–230 kg.

Pigs   Weaning to slaughter: porker, 93.6 kg intake and 44.8 kg carcase; baconer, 174.8 kg intake and 67.3 kg carcase. Again assuming a diet of barley (at 3239 kg ha^{-1}) 35 porkers per hectare and 18.5 baconers per hectare.

Broiler chicken   DM intake, 4 kg per bird. Carcase output, 1.45 kg. Assuming all-barley diet.

ruminants are probably more efficient, even if the problems of actually using such land (e.g. harvesting) by birds are disregarded.

Intensive egg production (meaning primarily intensive per unit of space) tends to be capital intensive as well and to use relatively little labour, with a fairly uniform requirement (Table 22.6). One result is that support energy is used with rather low efficiency (Table 22.7).

Water is not required in great quantity, by comparison with other forms of animal production, even for ducks and geese, unless they are kept in ponds (but this is not essential).

TABLE 22.6 Labour requirement for intensive egg production (after Nix, 1976)

Laying hens per worker	Labour hours per month per 100 hens	Total
3 000	7.5	225
5 000	5.0	250
8 000	3.25	260
12 000[a]	2.25	270

[a]Automatic egg collection

TABLE 22.7 Support energy use in egg production[a] (from Spedding and Bather, 1975).

Support energy inputs	Support energy inputs (MJ per bird)	Percentage of total support energy
[b]Feed (54.71 kg at 15.47 MJ support energy per kg)	846.5	80.6
Electricity		
incubation and hatching	4.43	0.42
brooder	20.30	1.93
controlled lighting	6.66	0.63
ventilation	63.41	6.04
automatic feeding	4.32	0.41
Total electricity	117.1	11.2
Capital		
buildings + equipment	44.0	4.2
field machinery (e.g. tractor)	6.6	0.6

TABLE 22.7 (*cont.*)

Veterinary + medical care	17.5	1.7
Replacement	5.8	0.6
Tractor fuel (feed transport + waste disposal)	10.1	1.0
Water	1.7	0.2
Litter	0.6	0.1
TOTAL ENERGY INPUT	1049.9	

Outputs

240 eggs at 0.051 kg edible egg at 6.62 MJ food energy/kg	12.24 kg 81.09 MJ	
1 cull carcase at 2.70 kg lwt (allowance for 12% mortality)	2.38 kg	
at 54% edible	1.29 kg	
at 6.03 MJ food energy/kg	7.78 MJ	
TOTAL FOOD ENERGY OUTPUT	88.81 MJ	

$$E\left(\frac{\text{energy out}}{\text{energy in}}\right) = \qquad 0.85$$

Support energy in:
[a]18-week rearing period + 52-week laying period, one laying period only: allowance has been made for mortality.
[b]This includes the support energy involved in both growing and processing the feed.

Factors Affecting Efficiency

The major factors that influence the biological efficiency of egg production by birds are included in Fig. 22.3, mainly in relation to output of eggs. Whether this is judged on a protein or energy basis, the dominant factor is the total weight of egg produced per unit of female body weight per unit of time. Once this ratio is very high, the effects of longevity, for example, rapidly decline in importance. To take account of the marked effect where longevity is very low, the unit of time considered needs to be of the order of at least one year.

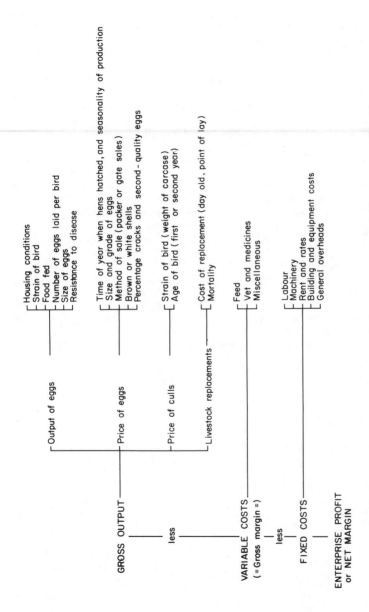

FIG. 22.3 Factors affecting the profitability of egg production (from Norman and Coote, 1971)

GROSS OUTPUT
— Output of eggs
    Housing conditions
    Strain of bird
    Food fed
    Number of eggs laid per bird
    Size of eggs
    Resistance to disease

— Price of eggs
    Time of year when hens hatched, and seasonality of production
    Size and grade of eggs
    Method of sale (packer or gate sales)
    Brown or white shells
    Percentage cracks and second-quality eggs

— Price of culls
    Strain of bird (weight of carcase)
    Age of bird (first or second year)

— Livestock replacements
    Cost of replacement (day old, point of lay)
    Mortality

less

VARIABLE COSTS
(= Gross margin =)
    Feed
    Vet and medicines
    Miscellaneous

less

FIXED COSTS
    Labour
    Machinery
    Rent and rates
    Building and equipment costs
    General overheads

ENTERPRISE PROFIT
or NET MARGIN

References

Baker, R. D., Large, R. V. and Spedding, C. R. W. (1973). Size of animal in relation to productivity with special reference to the ruminant — economic aspects. *Proc. Br. Soc. Anim. Prod.* **2**, 35–42.

Bolton, W. and Blair, R. (1974). Poultry Nutrition. M.A.F.F. Bull. 174. H.M.S.O., London.

Camp, S., Bourne, W. R. and Saunders, D. (1974). "The Sea Birds of Britain and Ireland", Collins, London.

Cott, H. B. (1954). Palatability of the eggs of birds. *Proc. Zool. Soc., Lond.* **124(2)**, 335–463.

Furness, R. W. (1978). Energy requirements of seabird communities: a bioenergetics model. *J. Anim. Ecol.* **47**, 39–53.

Lack, D. (1968). "Ecological Adaptations for Breeding in Birds", Methuen, London.

Murton, R. K. and Westwood, N. J. (1977). Birds as pests. *In* "Applied Biology", (Ed. T. H. Coaker), Academic Press, London and New York.

Nix, J. (1976). Farm Management Pocketbook. 7th edn. Wye College, Kent.

Norman, L. and Coote, R. B. (1971). "The Farm Business", Longmans, London.

Ricklefs, R. E. (1977). Composition of eggs of several bird species. *The Auk* **94**, 350–356.

Spedding, C. R. W. (1976). The relevance of various measures of efficiency. *In* "Meat Animals: Growth and Productivity", (Eds D. Lister, D. N. Rhodes, V. R. Fowler and M. F. Fuller), 29–41, Plenum Press, New York.

Spedding, C. R. W. and Bather, D. M. (1975). The importance of relative efficiencies in animal production. Paper presented at the Poultry Breeders Round Table Conference, Chester 1975.

Steel, T. (1965). "The Life and Death of St Kilda", The National Trust for Scotland, Edinburgh.

23

MILK AND MILK PRODUCTS

Milk production differs from most other animal production processes in several important ways.

It is a feature of mammals and is therefore always associated with warm-blooded animals, with all the implications discussed in the last three chapters. However, it does not share with meat production the property of adding to the daily maintenance cost, because it can be, indeed has to be, removed at frequent intervals — at least once a day and sometimes as often as three times daily. This makes it more comparable to egg production, which is why egg production can be regarded as, in a sense, intermediate between meat and milk production.

It is, of course, a feature of female mammals only and, whilst it is perfectly possible only to keep females for productive purpose (with the males only required for breeding), it is not yet possible to control the sex of the calves born to the dairy females that are kept. In order to ensure lactation, females have to be mated, generally at about annual intervals, and some 50% of the calves born are male (see Table 23.1 for sex ratios of different agricultural animals). Whereas in egg production the sex of the egg does not matter, since it is the final product, male calves are either not wanted at all in dairy herds or only wanted in very small numbers and from particular females. Increasingly, it is recognized that the surplus males are perfectly suitable for meat production and that if all the calves were female they would still not be needed in such numbers for replacement purposes.

Milk production therefore always has a major by-product and is linked with meat production. The relative importance of milk and meat in dairy production varies with the price of each, reflecting demand, supply, subsidies, theories about diet in relation to human health and availability and cost of feedstuffs.

The effects of these interactions on biological efficiency will be examined later in this chapter.

TABLE 23.1   Sex ratios of
different agricultural animals

	% males	Sources
Sheep	49–50	(1)[a]
Cattle	52	(1)
Pigs	50	(1)
Goats	57	(1)
Hens	48	(2)

[a]Sources: (1) Asdell (1964); (2)
Nalbandov (1958)

The first thing to recognize is the surprising number of species used for milk production in the world (Table 23.2): an enormous number of different breeds are also used within cattle, sheep and goats, although this is no greater than the number used for meat production. Even so, it means that the important physical and biological attributes of milk-producing animals cover a very wide range (Tables 23.3 and 23.4). However, it is worth noting that the contribution of sheep and goats to total world production of milk is only about 3.4% (Gall, 1975).

TABLE 23.2   Species used for
milk production

Buffalo
Camel
Cattle
Ass
Horse
Goat
Reindeer
Sheep
Yak
Eland

The product itself is remarkably uniform in composition between breeds of one species but differs considerably between species (Table 23.5) (Jenness and Sloan, 1970). Nevertheless, the basic process of milk secretion is similar between species (Linzell, 1972), as is the yield per unit weight of secretory tissue (Table 23.6).

TABLE 23.3 Attributes of species used for milk production

	Mature size Live wt (kg) ♂	♀	♂/♀ ratio for breeding	Annual reproductive rate	Annual yield (litres) Normal	Maximum
Buffalo	665–718	509–548	1:50	0.6	900–1 364	2 728–4 500
Camel	450–840	595	1:10–70	0.5	1 350	3 600
Cattle	700–800	450–700	1:30–50	0.9	3 864	11 365
Horse	1 000	700–900	1:70–100	1	11–17 per day	
Goat	48–84	45–77	1:40–50	1–3	600	2 045
Sheep	30–150	20–100	1:30–40	1–2+	150–200	1 200
Eland	700	450–500			350	630

TABLE 23.4 Characteristics of milk-producing animals

1. Ability to withstand extremes of climate, from the camel in hot arid areas to reindeer in very cold.

2. Adaptability to altitude (yak, goat), wet conditions (water buffalo) and various food types.

3. Ability to utilize a wide variety of food types from very productive pasture to lichens and browse.

4. Mostly ruminants and therefore able to digest fibrous materials more efficiently.

5. Can survive on land not suitable for crop production or make use of crop by-products.

6. Progeny can be used for replacement animals to produce milk or meat.

7. Supply nutritious, easily-utilized food daily for long periods, therefore storage not always necessary. Excess can be converted fairly easily to cheese or yoghurt which can be stored.

TABLE 23.5 Milk composition[a] of different species and different breeds (after Jenness and Sloan, 1970)

Species	Total solids	Fat	Casein	Whey protein	Lactose	Ash
Horse	11.2	1.9	1.3	1.2	6.2	0.5
Donkey	11.7	1.4	1.0	1.0	7.4	0.5
Pig	18.8	6.8	2.8	2.0	5.5	—
Camel						
bactrian	15.0	5.4	2.9	1.0	5.1	0.7
dromedary	13.6	4.5	2.7	0.9	5.0	0.7
llama	16.2	2.4	6.2	1.1	6.0	—
Reindeer	33.1	16.9	11.5		2.8	—
Eland	21.8	9.8	5.9	0.8	3.9	1.1
Cow						
Bos taurus	12.7	3.7	2.8	0.6	4.8	0.7
Bos indicus	13.5	4.7	2.6	0.6	4.9	0.7
Yak	17.3	6.5	5.8		4.6	0.9
Water buffalo	17.2	7.4	3.2	0.6	4.8	0.8
Goat	13.2	4.5	2.5	0.4	4.1	0.8
Sheep	19.3	7.4	4.6	0.9	4.8	1.0
Human	12.4	3.8	0.4	0.6	7.0	0.2

[a]Grams per 100 g or grams per 100 ml; no attempt is made to distinguish these modes of expression. Some figures are rounded off.

TABLE 23.6 Milk yield per unit weight of secretary tissue for different species (after Linzell, 1972)

Animal	Milk yield (kg per kg tissue)
Cow (dairy)	2–2.5
Goat (dairy)	1.75–2.41
Pig	1.3

Although milk may be secreted daily for a considerable lactation period, the quantity produced is not constant. Characteristic lactation curves are shown in Fig. 23.1. It is obvious that an animal cannot be efficient at milk production when it is not producing any and it follows that most resources will not be efficiently used when milk yield is low. This is so for virtually all resources and, in general, efficiency rises with milk yield for a given female, and where females are of similar size, those that give most milk tend to be most efficient.

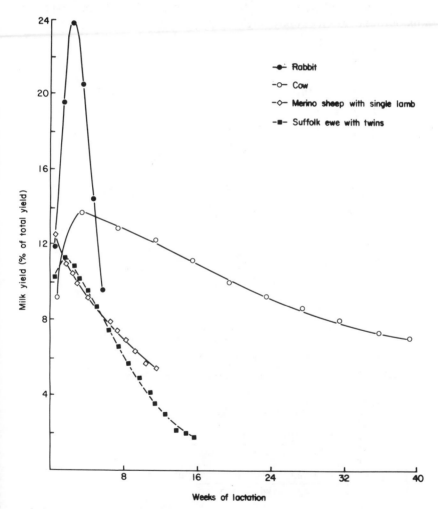

FIG. 23.1   Lactation curves of milk yield (g per day) expressed as a percentage of total milk yield (from Spedding, 1971)

This is true for the use of feed, although a particular feed may impose limitations, due to its nutritive value or its digestibility, on the milk yield it can sustain.

For an individual female, feed conversion efficiency varies with milk yield per day and is thus higher at the peak of lactation than at other times (deceptively so if liveweight changes occur). Between females of the same species and comparable size, feed conversion efficiency rises with yield per lactation (Fig. 23.2). Between species, compared at the peak of lactation and over a year, efficiency of feed conversion probably varies positively with milk output per unit of bodyweight.

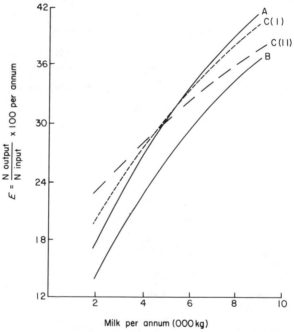

FIG. 23.2   Population efficiency of cattle (for milk) (from Large, 1973).

$$\text{Ratio} \quad A = \frac{\text{Primary output}}{\text{Input to dam} + \text{progeny}}$$

$$B = \frac{\text{Primary output}}{\text{Input to whole population}}$$

$$C = \frac{\text{Primary and secondary output}}{\text{Input to whole population}}$$

C(i) Spare calves valued at birth
C(ii) Spare calves taken to slaughter wt.

Energy and Protein Production

On good land, more energy and protein can usually be produced as milk than as meat (Table 23.7). The efficiency with which solar and support energy are used is generally higher for milk production (Tables 23.8 and 23.9), except for land of low primary productivity and for low fertilizer input, where the efficiency of support energy use may be much higher for meat production based on grazing.

TABLE 23.7   Energy and protein yields per unit of land from milk and meat

Product	Crude protein (kg per ha per yr)	Gross energy (MJ per ha per yr)	
Cows' milk	145	12 000	(a)
	118.5	8 770	(b)
Beef	31	2 400	(a)
	53–65	3 924–4 796	(b)
Sheep	22	2 500	(a)
	65	7 486	(b)
Pigs (bacon)	76	11 400	(a)
	105	14 438	(b)
Hen (broilers)	145	4 300	(a)
	135	7 071	(b)

Figures will vary according to the particular system used in the calculation. Those marked (a) are from White (1975), and those marked (b) have been calculated making the assumptions listed below. This illustrates the variation that can occur for UK conditions.

Assumptions for (b):

Cows' milk   Stocking rate of 0.8 cows per hectare for all food.

Beef   Suckler beef on grassland systems with stocking rate of 1.5 cows and calves per hectare for grazing and concentrates. Barley beef using calves from dairy herd.

Sheep   Intensive grassland system. Stocking rate of 17 ewes per hectare.

Pigs   Sows producing 20 progeny per year all fed on intensive diets.

Broiler poultry   Feed intake of broiler and share of parents' feed.

Where concentrate rations are used in the calculations the land area required to produce these feeds was calculated assuming average yields of crops.

TABLE 23.8 Efficiency of use of solar energy for milk production and other animal products

| Product | Output of protein (g) and energy (MJ) per MJ of solar energy received[a] | |
	Protein	Energy
Milk	0.004 4	0.000 36 (a)[b]
	0.003 6	0.000 27 (b)
Beef	0.000 9	0.000 07 (a)
	0.002 7	0.000 17 (b)
Sheep meat	0.000 7	0.000 08 (a)
	0.001 6	0.000 15 (b)
Pig meat (bacon)	0.002 3	0.000 35 (a)
	0.003 2	0.000 44 (b)
Hen eggs	0.003 4	0.000 18 (a)
	0.002 4	0.000 12 (b)

[a]Based on total annual radiation receipt of 33×10^6 MJ ha^{-1} yr^{-1}
[b](a) and (b) as for Table 23.7

The use of support energy has been examined in some detail for UK milk production and it can be seen from Table 23.10 where the main inputs occur. Comparable data for meat production are given in Table 23.11, to illustrate the chief differences between them, but it will be clear that much depends upon such factors as the level of fertilizer usage.

It should be remembered, of course, that the support energy inputs "downstream", involved in processing and distribution, are very high (Table 23.12) and even vary with such details as the method of bottling.

A great deal of water is used in all stages of milk production (Table 23.13) and a plentiful supply is essential for the process to operate efficiently at all.

Fertilizer inputs, provided that they increase herbage production and that this is harvested, either by grazing or cutting, increase the efficiency of land use and of the use of solar radiation (Fig. 23.3). Since individual animal efficiency is so dependent upon a high milk yield, it is very sensitive to grazing pressure and, where animals are fed substantially on pasture, correct stocking rates are of great importance. This is less so, of course, if a high proportion of the diet of the high-yielding animals comes from concentrates.

TABLE 23.9 Efficiency of use of support energy for milk production and other animal products

Product	Output of protein (g) and energy (MJ) per MJ of support energy used	
	Protein	Energy
Milk	8.53	0.71 (a)[a]
	7.30	0.54 (b)[b]
Beef	2.90	0.23 (a)
	1.66	0.11 (b)
Sheep meat	2.18	0.25 (a)
	2.50	0.23 (b)
Pig meat	4.22	0.63 (a)
Hen eggs	5.02	0.267 (a)
	1.50	0.077 (b)

[a]Calculated from White (1975)
[b]Calculated from data in Table 23.7: Spedding and Walsingham (1975); Spedding and Bather (1975)

TABLE 23.10 An example of support energy usage in UK milk production[a] (calculated from Downs, (1974)

	Energy usage (kJ per litre of milk)
Fertilizers	2 606
Concentrates	840
Hay	13
Straw	31
Tractors	5
Electricity	625
Fuel	571
Transport and processing of milk	422
Transport of resources to the farm	196
TOTAL	5 309

[a]The calculation is for a permanent pasture dairy farm of about 80 ha carrying on average 100 Friesian cows

TABLE 23.11 Support energy (kJ) used to produce 100 g edible meat (lamb) and milk

Input	Milk[a]	Lamb[b]
Fertilizers	253	3 090
Concentrates	82	1 930
Hay	1	—
Straw	3	—
Tractors	0.5	15
Electricity and fuel	116	2 139
Transport and processing of milk or lambs	41	471
Transport of inputs to farm	19	29
TOTAL	515.5	7 674

[a]Calculated from Downs (1974)
[b]Calculated from Table 25.11, Spedding and Hoxey (1975)

TABLE 23.12 Support energy inputs to a permanent pasture dairy farm[a] producing 5796 litres per ha per annum (calculated from Downs, 1974)

Support energy inputs	$(GJ\ ha^{-1})$
Upstream	10
On farm	4
Farm to retail	3.6 [b]
Retail to consumer	0.4 [c]
In the home	0.15

[a]Includes meat by-products of the dairy herd
[b]Includes bottling and the provision of glass bottles which are re-used 25 times
[c]Covers doorstep deliveries using electric vehicles

TABLE 23.13   Quantity of water used in modern milk production (data supplied by R. G. Kingwill, NIRD, 1978)

Daily requirements for 100 cows milked in 10/5 herringbone parlour	(litres per day)
Drinking	
Allow 60 litre per head per day (possibly more in summer and on dry winter rations)	6 000
Udder washing	
Basic 0.5 litre per cow per milking — double to allow for dirty cows	200
Milk cooling	
Assume milk tank	—
Cleaning milking plant	
Milking machine — say 40 litre × 5 units twice per day	400
Bulk tank — autoclean once per day	100
Cleaning parlour and collection yard — say	1 200
Total	7 900
Add 25% for variation, contingencies, etc. say	1 975
Grand Total	9 875
Required: 100 litres (22 gal) per cow per day	(60% for drinking)

Assuming no water required for cooling
(Note, where required this involves approximately 3 × volume of milk produced.)

Milk Products

There is no way that efficiency of human food production from any resource can be improved by the further processing of milk, where the latter can be drunk fresh by people to whom it is of value. The last point is not so for everyone: although milk is highly regarded as an economical food of high value in many countries (Rusoff, 1970), some people, indeed some races

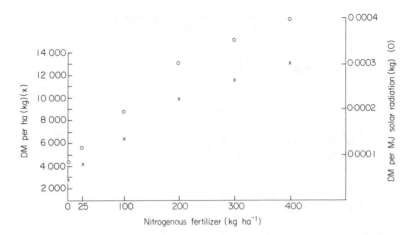

FIG. 23.3   Effect of fertilizer inputs on efficiency of land use for herbage production

(notably, many Asians), are unable to digest the lactose in milk (Rosens-weig, 1969), and others are allergic to some milk proteins. Whether it can be drunk fresh also depends on the supply matching the demand on a daily basis, or on an even shorter-term basis in very hot countries.

Not only does demand have to equal supply, and at a satisfactory price to both producer and consumer, but it has to do so in the same place, unless rapid distribution is possible. Thus milk produced in the Swiss Alps may have to be converted to a product, such as cheese, that will store without deterioration for the period required and be transportable, simply because the only practicable way to market it is for the farmer to carry it down to the valleys with him weeks or months later.

The form of the cheese made may be determined, or originally have been determined, by considerations of this kind, but cheeses made for sale will generally now be aimed at a particular market and will have acquired quite rigid standards of flavour and quality.

There is little point, therefore, in comparing the efficiency with which milk nutrients are converted into *different* cheeses: the general features are indicated in Table 23.14.

Milk and Meat Production

As mentioned at the beginning of this chapter, meat is bound to be a by-product of milk production, even when it is regarded as negligible (usually because unwanted calves are slaughtered at about one week old).

TABLE 23.14   General features
of the efficiency with which milk
nutrients are converted into
cheese (after Van Slyke and
Price, 1949)

Average amounts of fat, proteins
and water recovered in cheddar
cheese made from 100 lb
(45.4 kg) of milk

	(lb)	(kg)	(%)
Fat	3.39	(1.54)	91.1
Protein	2.37	(1.08)	75.7
Water	3.72	(1.69)	

$$\text{Percentage recovery} = \frac{\text{Wt of fat (or protein) in cheese}}{\text{Wt of fat (or protein) in milk}} \times 100$$

Where the efficiency of feeding the progeny for meat production is less than the whole milk production system without this additional enterprise, there is clearly an argument for terminating the life of the calf (or lamb, or kid) as soon as possible.

This may seem odd when beef is nevertheless being produced by someone, from calves that cost a great deal to produce.   However, calves are unlikely to be slaughtered at birth where a demand for them exists (and is expressed) and they can be sold. But they are not always in the right place at the right time and there are costs involved in marketing them and transporting them to where they are needed.

The effect on efficiency of feed conversion is illustrated in Table 23.15; meat from culled breeding females is also included, since this is likely to be a substantial source of meat (and income). The relative efficiency of producing meat by slaughtering the cow as a once-bred heifer was dealt with in Chapter 20.

If the progeny are to be reared at all, whether for meat or replacement dairy females, some of the milk produced may be required for rearing. Virtually the whole range of possibilities is practised in one part of the world or another. The calf, kid or lamb may receive none of the milk produced, except the colostrum. Such progeny may be slaughtered, or reared on milk substitutes. There are many instances where calves are fed on milk substitutes, or even dried cows' milk, whilst their own cow's milk is being sold for

TABLE 23.15   Feed conversion efficiency of a dairy herd[a] with calves slaughtered at birth or at beef weights

	Inputs (kg DM)		Outputs (kg)	(kg)
Feed for cows	422 000	Milk	562 200	
		25 cull cows (303 kg)		7 575
		60 calf carcases (24 kg)		1 440
(1) Total feed	422 000	Total carcase output		9 015
Feed for calves taken to beef wt	102 000	60 beef carcases (225 kg)		13 500
(2) Total feed	524 000	Total carcase output		21 075

Carcase output per kg feed DM: (1) = 0.02; (2) = 0.04

[a]Theoretical herd of 100 milking cows yielding 1200 gal (5622 kg) average, producing 90 calves per year, 20 of these are required for replacements and 5 for mortality losses, leaving 60 calves for immediate slaughter or taking on to beef. 25 cows to be culled

human consumption. This may be due to pricing arrangements or simply because the calves are sold to be reared in quite separate enterprises.

In Spain, when the price of ewes' milk has been high (for cheese making), it has made economic sense to remove the lambs and rear them on milk substitute (made in another country and transported) or even on dried cows' milk.

At the other extreme, most of the milk produced may be consumed by the progeny and only a small proportion removed to be drunk by the owner.

In between these extremes are all ages of weaning, from a few weeks to several months, with the progeny being transferred to solid feeds. Very early weaning usually reduces the growth rate of the progeny, at least initially, but encourages early development of the rumen and increases the proportion of the total diet that is fibrous and usually, therefore, cheap.

The effect of early weaning on efficiency of feed conversion to meat has been studied on sheep (Table 23.16) but, as with the conversion of milk into cheese and butter, conversion to meat must involve further losses.

Factors Affecting the Efficiency of Milk Production

These are shown in Fig. 23.4 and the response to variations in milk yield have already been illustrated (Fig. 23.2).

TABLE 23.16   The effect of early weaning on the feed conversion efficiency of sheep (from Walsingham *et al.*, 1975).

Values of E^a measured over 215 days for three breed crosses weaned at three ages. (Values are for three ewe and lamb units, except where indicated otherwise in parentheses.)

Breed cross		Age of twins at weaning (weeks *post partum*)					
♂	♀	4		10		16	
		Mean	SE	Mean	SE	Mean	SE
Suffolk	Welsh M.	11.25	0.65	10.60	0.43	10.18	0.24 (2)
Welsh M.	Welsh M.	9.51	0.07 (2)	—	—	9.35	0.19 (2)
Kerry Hill	Kerry Hill	8.98	0.43	10.18	0.73	10.59	0.33

$$^aE = \frac{\text{Wt. of carcase output (kg)}}{\text{Digestible organic matter consumed by ewe and lamb (kg)}} \times 100$$

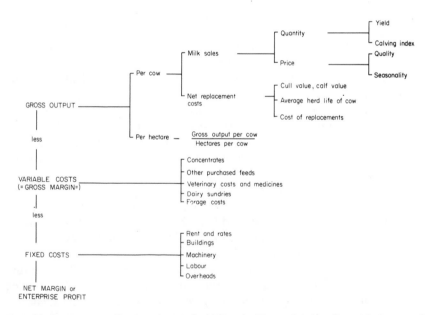

FIG. 23.4   Factors affecting the profitability of milk production (from Norman and Coote, 1971).

Milk production has been highly mechanized in most developed countries but is nevertheless very dependent on labour being available at exactly the right times. Furthermore, in milk production almost more than any other animal production enterprise, the behaviour of the labour force may exert a bigger influence on productivity than the behaviour of the animals. It is, of course, easy to imagine how disruptive behaviour could deleteriously affect other animal enterprises, perhaps especially those involving housed birds, but there appear to be more subtle influences at work in the case of milkers and their cows. Amongst the biological factors, conception rate is of major consequence and herd fertility must be counted as a major influence in the productivity of dairy cows (Table 23.17).

TABLE 23.17 Productivity of dairy cows in relation to conception rates (from James and Esslemont, 1979)

	Effect of increasing calving interval by one day, from 380 days, on annual milk yield											
Month of first calving	Jan	Feb	Mar	Apr	May	June	July	Aug	Sept	Oct	Nov	Dec
Reduction of annual milk yield (litre)	13.65	12.74	9.55	10.92	10.92	0.91	5.00	5.46	0.91	11.83	20.02	9.55

In terms of strict milk production, the major factors are bound to be those that determine the annual milk yield and, calculated over a long enough period, this ultimately includes longevity and disease incidence as well as individual lactation yields. The resources required to produce 1 kg of milk (or milk energy or protein) will also be influenced by the environment, animal size and species, methods of management and the quality and quantity of feed available.

It is not always possible, however, to distinguish resources used for milk production and those used in the production of inevitable by-products, such as progeny, body tissue, wool and hair. In a sheep that is being milked, growing wool and changing in body weight, for example, it is not possible to calculate wholly independent efficiency ratios. Nor is there any completely satisfactory way of expressing all these in common terms, and the literature on efficiency of milk production contains many different expressions using different units (Moe and Tyrrell, 1975). Money is not necessarily adequate, especially for body weight changes, but neither energy nor protein give comparable weightings to wool, milk and body fat.

Of course, this is no different in principle from the fact that any calculation of efficiency related to one product only ignores all other inputs (and faeces, for example, are always an output). This is particularly so where it is difficult to distinguish a major product or where it is actually intended to focus on the by-products.

References

Asdell, S. A. (1964). "Patterns of Mammalian Reproduction", 2nd edn, Constable, London.

Downes, D. M. (1974). Support energy in milk production. B.Sc. Honours Dissertation. Department of Agriculture and Horticulture, University of Reading.

Gall, C. (1975). Milk production from sheep and goats. *Wld Anim. Rev.* No. 13, 1–8.

James, A. D. and Esslemont, R. J. (1979). The economics of calving intervals. *Anim. Prod.* **29**, 157–162.

Jenness, R. and Sloan, R. E. (1970). The composition of milks of various species: a review. Review Article No. 158, *Dairy Sci. Abstr.* **32(10)**, 599–612.

Large, R. V. (1973). Factors affecting the efficiency of protein production by populations of animals. *In* "The Biological Efficiency of Protein Production", (Ed. J. G. W. Jones), 183–199. Cambridge University Press, Cambridge.

Linzell, J. L. (1972). Milk yield, energy loss in milk and mammary gland weight in different species. Review Article No. 169 *Dairy Sci. Abstr.* **34(5)**, 351–360.

Moe, P. W. and Tyrrell, H. F. (1975). Efficiency of conversion of digested energy to milk. *J. Dairy Sci.*, **58**, 602–610.

Nalbandov, A. V. (1958). "Reproductive Physiology", W. H. Freeman, San Francisco.

Norman, L. and Coote, R. B. (1971). "The Farm Business", Longmans, London.

Rosensweig, N. S. (1969). Adult human milk intolerance and intestinal lactase deficiency. A review. *J. Dairy Sci.* **57(5)**, 585–587.

Rusoff, L. L. (1970). Milk: its nutritional value at a low cost for people of all ages. *J. Dairy Sci.* **53(9)**, 1296–1302.

Spedding, C. R. W. (1971). "Grassland Ecology", Oxford University Press, Oxford.

Spedding, C. R. W. and Bather, D. M. (1975). The importance of relative efficiencies in animal production. Paper presented to Poultry Breeders Round Table Conf.

Spedding, C. R. W. and Hoxey, A. M. (1975). The potential for conventional meat animals. *In* "Meat", (Eds D. J. A. Cole and R. A. Lawrie), Butterworths, London.

Spedding, C. R. W. and Walsingham, J. M. (1975). Energy use in agricultural systems. *Span* **18**, 7–9.

Van Slyke, L. L. and Price, W. V. (1949). "Cheese", Routledge and Kegan Paul, London.

Walsingham, J. M., Large, R. V. and Newton, J. E. (1975). The effect of time of weaning on the biological efficiency of meat production in sheep. *Anim. Prod.* **20**, 233–241.

White, D. J. (1975). Energy in agricultural systems. Paper presented to Ann. Conf. Inst. Agric. Engineers, The Use of Energy in Agriculture. London 1975.

24

OTHER FOOD PRODUCTS

There are very few human foods derived from agricultural animals that do not fit under any of the previously-used headings — meat, fish, eggs and milk. A possibility for the future would be the large scale culture of invertebrates, but this is not yet a major agricultural activity. Probably it would be misleading to describe the product of such culture under any of these main headings and a new category might be required. However, it seems more likely that direct human needs could be more efficiently met by plant micro-organisms and that the role of invertebrates would be to convert wastes into a feed suitable for pigs and poultry.

Currently, only two additional categories need be mentioned: (a) secretions other than milk and (b) by-products of meat and fish production.

Secretions Other than Milk

Only two are of any significance — blood and honey.

Blood

The practice of drinking blood harvested from live animals appears to be confined to certain African tribes (e.g. Masai, Pokot, Dinka, Nuar, Karimojong) and their animals and is of declining importance. It has many advantages in nomadic circumstances, particularly where nutrition is poor.

As described for milk production, it allows frequent removal of a product, production of which does not add to the maintenance burden of the animal. Presumably, blood can be removed at much less frequent intervals than milk, without causing production to stop, and is thus relevant to very poor nutritional conditions.

It is not as hygienic a process as milk production, because of the presence

314

of blood parasites, for example, but appears to lead to few problems where it is practised. There is not a great deal of quantified information on the subject (Table 24.1) and too little to warrant discussion of efficiency. It is perhaps worth remembering that, although the practice does not appeal to most people, the blood of hunted animals is often drunk, that blood is a normal constituent of meat and that it is regularly used as a processed by-product of meat production.

TABLE 24.1   The use of blood from live animals

Tribe	Amount of blood collected at one time (litres)	Frequency of collection (No. per yr)	Sources
Dinka (Sudan)			
Cattle	2–3	2–4	(1)
Pokot (Kenya)			
Camels	5–6	3–4	
Bulls and oxen	2	up to 6	(2)
Cows	1–2	3–4	
Sheep and goats	0.7	1–2	
Karimojong (Uganda)			
Cattle	2–4.5	2–4	(3)

Sources: (1) Payne and El Amin (1977); (2) Brown (1979); (3) Dyson-Hudson and Dyson-Hudson (1969)

Honey

Before the development of sugar cane and sugar beet, honey must have been a major source of sugar for sweetening. It has remarkable properties still, not all of them well understood, for long-term storage and of a dietary nature.

As a commercial production process it is all based on the honey-bee (*Apis mellifera*), although this exists in at least 14 varieties.

The production characteristics of the honey-bee are given in Table 24.2, from which it will be seen that remarkable quantities of honey can be produced, especially where queens have been subjected to some selective breeding and the supply of nectar is good.

The efficiency of honey production is unusually hard to define in useful terms, partly because it is never the sole use of the most important resources

BIOLOGICAL EFFICIENCY IN AGRICULTURE

TABLE 24.2   The production characteristics of the honey bee

Type of beekeeping		Honey harvested (kg per hive per yr)	Total honey equiv.[a] of feed collected by colony (kg per hive per yr)	Honey collected (g per bee per yr)[b]	Honey collected (kg per ha per yr)[c]
Static					
average output	UK	15–20	150–200	0.75–1.0	3.6–4.8
best output	Australia	100	286	1.43	22.8
Migratory					
average output	Australia	200	570	2.85	45.3
record output	Australia	350	1 000	5.0	79.6

Assumptions

[a] Honey harvested in UK and Australia is approximately 10% and 35% respectively of total produced by the bees.
[b] 200 000 worker bees per colony per year.
[c] Apiary of 30 hives (UK); 100 hives (Australia); flight range of 2 km from the hives i.e. 1257 ha. The values shown compare with the theoretical honey potential of plants having the highest yields of Honey of over 500 kg ha^{-1}.

Source: Crane (1975)

(e.g. land, flowers) and partly because the significance of the bees as pollinators of agricultural crops is very difficult to assess (see Chapter 26).

Equipment, particularly the hives themselves, has remained fairly simple, and hives are usually small and separate. There is, of course, an obvious limit to the number of colonies that it would ever be sensible to place at one site, since there is no sense in forcing bees to forage over unnecessary distances. This aspect of efficiency can be readily calculated (Tables 24.3 and 24.4) and there may be no advantage in siting more than 30–100 hives together, in permanent sites, varying with the availability of nectar. The main advantages of placing hives together is to the beekeeper, in order to reduce labour requirements.

TABLE 24.3 The effect of foraging distance upon colony net gains of honey (from Ribbands, 1952)

		Mean gain per colony (kg)		
	Crop and date	At crop	⅜ mile away	¾ mile away
Apple	(1949, 2–12 May)	6.0 ± 1.2	4.8 ± 1.2	0.9 ± 0.6
	(1950, 28 April–18 May)	1.4 ± 1.0	–2.0 ± 0.7	–5.5 ± 0.5
Lime	(1949, 28 June–15 July)	25.4 ± 2.7	24.5 ± 4.3	14.8 ± 1.9
	(1950, 1–15 July)	16.0 ± 2.8	11.0 ± 1.5	9.1 ± 2.2
Heather	(1949, 4 Aug.–20 Sept.)	27.1 ± 2.6		23.2 ± 1.0
	(1950, 2 Aug.–5 Oct.)	11.6 ± 0.5		1.1 ± 0.3
Cabbage	(1949, 28 April.–2 May)	–0.2 ± 0.3	–2.2 ± 0.05	–1.3 ± 0.1
Onion	(1949, 7–30 July)	17.2 ± 2.6		13.8 ± 3.4
	Total gain per colony, all experiments together:			
	1949	75.5		51.4
	1950	29.0		4.7

TABLE 24.4 Calculated effect of foraging distance on the weight of nectar brought into the hive (Ribbands, 1952)

Time spent on crop and in unloading (min)						
	5	15	30	60	120	240
Reduction of nectar intake at ⅜ mile (%)	39	18	10	5	3	1
Reduction of nectar intake at ¾ mile (%)	56	30	18	10	5	3

Although bees are classed as cold-blooded, the colony is quite capable of maintaining its temperature well above ambient, even outside the hive. There is thus an appreciable energy cost in maintaining a colony round the year and additional energy (as sugar syrup) is normally supplied in the autumn (in temperate countries), where the removal of honey has left insufficient for the needs of the colony.

The energy exchanges in honey production were illustrated in Table 24.2. Only a small proportion of the energy available in the nectar of flowers actually finishes up as honey for human consumption. A moment's reflection on alternative ways of harvesting it, especially where the flowers are wanted and have to be pollinated, suggests that they would be extremely difficult to devise at all, quite apart from any possibility of being energetically more efficient than the bee.

By-products of Meat and Fish Production

There are by-products, such as faeces (see Chapter 27), of all forms of animal production, and meat is one of the major by-products of both milk and egg production. By-products for human food that are not themselves covered under meat, fish, egg or milk production, come chiefly from meat and fish production systems.

The enormous variety of uses to which different parts of the slaughtered animal are put are illustrated in Table 24.5 for cattle and similar considerations apply to all the major animal species used in agriculture (M.L.C., 1975). Use is thus already made of most parts of the animal body and some of them are quite important sources of human food. Exactly which part of the animal is eaten, or preferred, varies enormously with locality and custom. It is rather extraordinary that some items (e.g. brains and eyeballs) should be regarded as delicacies by some and with revulsion by others. This points to the difference that can occur between agricultural production, often loosely thought of as food, and what is recognized as edible food by the consumer (Tudge, 1977).

The efficiency with which by-products are produced can only have a rather restricted kind of meaning, with a very limited use and significance.

After all, no-one is deliberately setting out to produce them and they would be produced anyway, by definition, as part of the main process. It is possible to have production processes, as in some systems of sheep production, where two products are of about equal value (e.g. meat and wool) and it would always be possible to apportion the resources used in proportion to the quantity or value of the products produced.

For many purposes, the only satisfactory answer is to consider production, of all forms, from the entire system and weight the products in some

TABLE 24.5 Utilization of parts of the slaughtered animal other than the carcase

Use	Examples of parts used
Food	Liver, tongue, ox-tail, kidney
Animal feed	Spleen, blood, stomach
Pharmaceuticals	Liver, intestines (heparin), gall, pancreas (insulin)
Fertilizer	Blood, bone
Glue	Bones
Fire-suppressing foam	Blood
Rennet	Abomasum of suckling calves
Neatsfoot oil	Feet
Buttons and handles	Bones
Gelatine	Bones
Tallow	Fat
Edible fats	Internal fats and carcase trimmings
Leather	Hide
Sausage casings	Small intestines
Surgical ligatures	Small intestines
Paints	Gall
Dyes	Gall

appropriate manner (which might be energy or protein or units of human food or money).

Within a production system, it may be desirable to make comparisons of by-product output: this is most usefully done per animal or per unit of land and makes independent sense provided that the output of the main product remains constant. Where this also varies, both have to be considered together and, to do this, they have to be expressed in common terms.

The only certain message is that at some point a complete assessment of a production system has to be undertaken and must include all outputs and all inputs. It may then be found that the contribution of the by-products is crucial to the economic success of the enterprise.

References

Brown, J. (1979). Personal communication.

Crane, E. (1975). "Honey", Heinemann, London.

Dyson-Hudson, R. and Dyson-Hudson, N. (1969). Subsistence herding in Uganda. *Scientific Am.* **220(2)**, 76–89.

M.L.C. (1975). "The Market for Animal By-products". M.L.C. Technical Bulletin 21.

Payne, W. J. A. and El Amin, F. M. (1977). An interim report on the Dinka livestock industry in the Jonglei area. Tech. Rep. No. 5. The Democratic Republic of the Sudan, Economic and Social Research Council N.C.R.

Ribbands, C. R. (1952). The relation between the foraging range of honeybees and their honey production. *Bee World* **33**, 2–6.

Tudge, C. (1977). An end to meat mythology. *Food Policy* **2(1)**, 82–85.

25

NON-FOOD ANIMAL PRODUCTS

It is often difficult to draw a satisfactory line between main products and by-products and their relative importance can quite easily be reversed if economic circumstances change markedly. Thus sheep may nearly always be regarded as producing meat and wool (there are few breeds that produce no wool at all) and which of these is regarded as the main product really depends on their relative prices.

There are, however, a number of important non-food products that are main products in their own right (or could be so) or are major by-products.

All the principal agricultural animals possess skins, usually with an integument of hair, wool or feathers, and these naturally become available at slaughter. The proportion of body weight is often appreciable (Table 25.1) and the total world production of hides and skins is very large (273.6, 386.4 and 145.8 million pieces of cattle and calfskin, sheepskins and goatskins respectively in 1975; F.A.O., 1976).

TABLE 25.1   Hide or skin as a proportion of body weight (taken from the literature)

	Hide wt as % of slaughter wt	Hide wt as % of empty body wt (EBW)
Steers	5.3– 5.8	9–10
Bulls	7.1– 7.3	
Cattle	7.2– 9.1	
Lambs		
(skin + wool)	6.9–10.8	
(41 wks + wool)		10
Pig (16 weeks)		2

321

Only fur-producing animals (Table 25.2) with very valuable pelts are likely to be kept for these alone and, in these cases, the value of the output is so great that biological aspects of efficiency count for little. That is why it is even possible to farm carnivorous animals for fur production.

In general, the major non-food products are of a kind that can be harvested from the live animal or, in some sense, independently of it. The chief examples of the former are wool and mohair and, of the latter category, secretions such as silk. There are some other secretions of importance (snake venom, semen) but silk is quantitatively the most important.

TABLE 25.2   Animals kept for their pelts alone

Mink
Silver fox
Chinchilla
Marten

Source: F.A.O. (1970)

Wool and Hair

Several kinds of animal produce fibrous coats which are used by man (Table 25.3) but sheep are by far the most important.

TABLE 25.3   Animals producing fibrous coats which are used by man

Sheep
Llama
Alpaca
Angora goat
Cashmere goat
Bactrian camel
Angora rabbit

Source: Commonwealth Secretariat (1973)

Wool

This is an excellent example of a product that is very hard to express in quantitative terms that are of the slightest use for comparative purposes. There are measures of quality, concerned with the length and fineness of the fibres, that are related to the insulating properties of wool or to attributes that influence its manufacturing potential. But the *weight* of wool is of limited use when quality differs, and the chemical content may give little indication of the insulating properties of wool.

The insulating and rain-shedding (and protection against excessive insolation) properties of a fleece matter, of course, to the the sheep and thus to the producer, but the properties that concern the manufacturer may be somewhat different.

This is also a good illustration of the extent to which price may primarily reflect demand rather than a value judgement as such. For different purposes, such as making shirts or making carpets, one type of wool may be better than another, but the fact that it is "better" does not necessarily make it more valuable. Its monetary value is determined by how much is wanted and how much is available. Of course, the costs of production also bear on this question.

Most wool is produced from sheep in fairly extensive systems, utilizing relatively cheap land of low primary productivity. Sheep occupy a wide range of environments from arid areas in the Middle East and Australia to wet, cold mountains in Scotland and Iceland. The type of wool, and thus the breed of sheep, is often related to the nature of the environment and there is usually little to be derived from comparison of such totally different situations.

There are, however, several very important aspects of biological efficiency that are characteristic of wool production in general. Like hair, wool continues to grow even when an animal is losing weight: it is therefore much less dependent than meat production on the adequacy of current nutrition. This makes it a suitable form of production in environments with a poor feed supply and with marked seasonality.

One major difference between wool production and that of meat, eggs, fish and milk, is that it is related to the surface area of the animal rather than to its weight or volume. So, whereas a high surface area/volume ratio tends to be disadvantageous to warm-blooded animals engaged in most forms of production, this is not so for wool-producing animals. These disadvantages have to do with the fact that heat loss and maintenance requirement are related to surface area, while performance is more likely to be related to weight (or volume of biological tissue). For otherwise similar animals, therefore, more wool tends to be produced by those with a large surface area. For this reason, the Merino sheep was selectively bred for large surface area, involving a massive development of skin folds (with considerable disadvantages in terms of shearing and the harbouring of parasites).

One result of all this is that wool production tends to be less sensitive to stocking rate than other forms of production, in terms of output per animal (see Table 25.4 and Fig. 25.1). Furthermore, the efficiency of feed conversion into wool has been found to be independent of the size of the sheep (Coop and Hayman, 1962; Ryder and Stephenson, 1968). The same quantity of feed should therefore produce the same amount of wool, whether it is

TABLE 25.4　The relationship between wool output per sheep and stocking rate

Stocking rate (sheep per ha)	Wool output (clean) (kg/sheep)				
	1958–59[a]	1958–59[b]	1959–60[b]	1960–61[b]	1961–62[c]
4.8	3.8 ± 0.18	3.8 ± 0.18	4.5 ± 0.18	3.6 ± 0.18	3.5 ± 0.18
9.7	3.1 ± 0.14	3.5 ± 0.14	4.8 ± 0.18	4.1 ± 0.14	3.5 ± 0.14
14.5	3.0 ± 0.14	3.3 ± 0.14	4.7 ± 0.23	4.4 ± 0.18	3.2 ± 0.18
21.8			4.5 ± 0.18	3.9 ± 0.14	3.2 ± 0.14

[a]Adult sheep on annual pasture
[b]Adult sheep on perennial pasture
[c]Weaner sheep on perennial pasture

Calculated from Arnold *et al.* (1964)

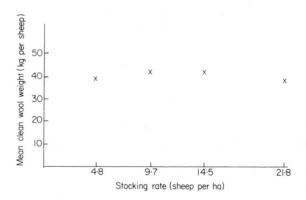

FIG. 25.1　Effect of stocking rate and wool production per sheep (after Arnold *et al.*, 1964)

fed to one large sheep or to two small ones. Clearly, this can only be true within limits, since sheep fed less than they need for maintenance will eventually die. However, the comparable proposition for meat production from warm-blooded animals is not true at all.

Part of the explanation might be that wool production does not, of itself, require a great deal of energy: certainly the energy content of the product has little relevance and this makes energetic efficiency calculations of rather limited use. The efficiency of protein production as wool, on the other hand, looks high (Table 25.5) but this assessment also adds little of value.

TABLE 25.5 Efficiency (E) of protein production by sheep (calculated from Large, 1973)

No. of lambs per annum	E^a	
	A	B
1	0.093	0.056
2	0.125	0.086
3	0.145	0.110
4	0.160	0.125
5	0.172	0.138
6	0.180	0.146

$$^a E = \frac{\text{N output (A with and B without wool)}}{\text{N input to the sheep population per yr}}$$

Relations between efficiency, animal size and stocking rate are of great practical significance and also of theoretical interest, but most of the useful efficiency ratios for wool simply express it as dry weight. This expression has to be used with some care, since natural wool contains considerable amounts of sweat and grease (Table 25.6), which greatly influence energy and protein content.

The use of support energy is generally low in raw wool *production*, in marked contrast to that used in the production of artificial fibres (see Table 14.11, p. 180). However, if the support energy subsequently required for industrial processing is included, the difference is of much less significance.

For most of the production processes discussed in this book a nutritionally balanced diet has been assumed. Wherever one component is grossly deficient, however, responses in productive efficiency to additions of the missing element are bound to be substantial. A good example, indeed a spectacular one, is provided by additions of sulphur-containing amino acids. Wool growth is very sensitive to these particular amino acids and they may be destroyed in the rumen even if they are present in the diet. If the dietary proteins are protected by treatment with formalin (or even by heating), digestion in the rumen may be avoided, more of the essential amino acids may be absorbed and significantly more wool grown (Table 25.7).

In principle, of course, the same is true of any limiting nutrient or any other limiting factor; efficiency will increase following additions until some other factor becomes limiting and prevents further response.

TABLE 25.6 Percentage of clean wool, wax, suint and dirt in greasy wool

Clean[a] wool	Wax	Suint	Dirt
60–64.5	14–21	4–6	10–16.5

[a]Clean wool containing approximately 10% water

After Roberts (1954) in Ryder and Stephenson (1968)

TABLE 25.7   Effect on wool growth of protecting dietary proteins with formalin (after Ferguson et al., 1967)

	Wool growth (g per head per day)		
Week	Control[a]	Experimental Group A[b]	Experimental Group B[c]
2	6.5	9.0	8.7
4	6.7	10.8	11.1
6	6.3	11.2	10.7
8	6.8	11.4	11.3
10	6.9	11.0	9.8
13	6.8	10.8	7.9
15	6.5	10.9	7.4
17	6.9	10.7	8.0

Diets

[a]Control: 800 g per day of a diet of wheat and lucerne chaff 50:50 for 12-week preliminary period and continued for 18-week experimental period.
[b]Group A: ration as for control group plus 80 g per day casein treated with formaldehyde for the experimental period.
[c]Group B ration as for Group A, until week 8 of experimental period; for the remainder of the period untreated casein was fed.

Silk

Silk is also largely proteinaceous, being a mixture of two proteins chiefly, fibroin and sericin. It is secreted by silk moth larvae when pupating and its recovery involves destruction of the pupae, which are commonly fed to fish.

There are several different species of silk moth, that most commonly used commercially being *Bombyx mori*, of which more than a thousand varieties exist (Yokoyama, 1963).

Many food plants are eaten by silkworms but Mulberry (*Morus alba*) leaves are usually provided.

Although silk is a secretion, its production is more akin to that of meat than, for example, wool or milk. It is only secreted at one time, so the materials for it are, in fact, stored over the life of the larvae in its own body. So it is harvested all at once at the end of the animal's life.

Feed consumption and silk production have been measured in considerable detail, so there are good figures for feed conversion efficiency (Table 25.8) but there is much less information for output per unit of land. Labour requirements are high and a considerable amount of equipment (cages, trays and so on) is needed.

TABLE 25.8 Feed conversion efficiency by *Bombyx mori* (after Waldbauer, 1968)

Temperature (°C)	Feed	Instar or stage	ECI[a]	ECD[b]
22	*Morus alba* Goshoerami variety (Moraceae)	Whole larval stage	23	62 (1)[c]
22	*Morus alba*	Instar IV–V	19	48 (2)
27	*Morus alba*	Instar IV	21	47 (3)

[a]ECI = Efficiency of conversion of ingested food to body substance

$$ECI = \frac{Wt\ gained\ (wet)}{Wt\ of\ feed\ ingested\ (DM)} \times 100$$

[b]ECD = Efficiency with which digested food is converted to body substance

$$ECD = \frac{Wt\ gained\ (wet)}{Wt\ of\ feed\ ingested\ (DM) - wt\ of\ faeces\ (DM)} \times 100$$

[c]Sources: (1) Hiratsuka (1920); (2) Shyamala *et al.* (1960); (3) Soo Hoo and Fraenkel (1966)

The use of support energy is less well documented and, in any case, will depend upon labour usage.

The biological efficiency of the silk moth, in terms of total production, is quite high, because the entire output can be used. The digestive efficiency of the larvae is similar to that of other herbivorous insects (see Table 25.9): the faeces can be used as a fertilizer. The larval bodies are, of course, entirely converted into pupal mass and this provides a useful feed for fish. The requirement for breeding adults is small, even though the sex ratio has to be about 1:1, because the females lay about 200 eggs each. Furthermore, the adults do not feed at all during their short lives (a few days) and their bodies can also be used as fish feed.

Not only is the egg output of female moths substantial but the eggs can be stored for several months. This is a feature of considerable importance in making efficient use of seasonally available feed. It is the most extreme

TABLE 25.9 Approximate digestibility of feed by *Bombyx mori* larvae and other herbivorous insects (after Waldbauer, 1968).

Insect, food and authority	Instar	
	IV	V
Bombyx mori		
Leaves of *Morus alba*	38	37
Hiratsuka (1920)		
Phalera bucephala		
Leaves of *Carpinus betula*		
Dry weight[a]	24	19
Fresh weight[a]	46	45
Evans (1939)		
Schistocerca gregaria		
Mixed grasses	34	35
Davey (1954)		
S. gregaria		
Leaves of *Ficus* sp.	38	31
Husain *et al.* (1946)		

All figures are on a dry weight basis unless otherwise indicated

[a]Means of approximate digestibility (AD) measured over 24 hours at intervals of several days, probably not representative of the AD for the entire instar

example of this amongst agricultural animals, since it would be possible to utilize fully a period of, say, six months of leaf production, without any feed cost at all during the other six months of the year, during which fertile eggs could be stored. Of course, as with hibernating animals, some loss in weight must occur during "storage" but, in the case of silk moth eggs, the loss is small and is not considered to reduce fertility appreciably. This is a general property of many insects. They have short adult lives and longer larval phases, and their eggs may survive for quite long periods. Other insects overwinter in the adult phase, at little energy cost, to produce numerous eggs and progeny the following spring.

Where the feed supply is only available for a very short period each year, this life-style is clearly the most efficient.

References

Arnold, G. W., McManus, W. R. and Bush, I. G. (1964). Studies in the wool production of grazing sheep. I. Seasonal variation in feed intake, liveweight and wool production. *Aust. J. Expl. Agric. Anim. Husb.* **4**, 392–403.

Commonwealth Secretariat (1973). "Industrial Fibres", Commonwealth Secretariat, London.

Coop, I. E. and Hayman, B. I. (1962). Liveweight–productivity relationships in sheep. II. Effect of liveweight on production and efficiency of production of lamb and wool. *N.Z. J. Agric. Res.* **5**, 265–277.

Davey, P. M. (1954). Quantities of food eaten by the desert locust, *Schistocerca gregaria* (Forsk.) in relation to growth. *Bull. Ent. Res.* **45**, 539–551.

Evans, A. C. (1939). The utilization of food by certain lepidopterous larvae. *Trans. R. Ent. Soc. Lond.* **89**, 13–22.

F.A.O. (1970). The world hides, skins, leather and footwear economy. Commodity Bull. Series 48. F.A.O., Rome.

F.A.O. (1976). F.A.O. Commodity Review and Outlook 1975–1976. F.A.O., Rome.

Ferguson, K. A., Hemsley, J. A. and Reis, P. J. (1967). Nutrition and wool growth. *Aust. J. Sci.* **30(6)**, 215–217.

Hiratsuka, E. (1920). Researches on the nutrition of the silk worm. *Bull. Ser. Expl. Sta. Japan* **1**, 257–315.

Husain, M. A., Mathur, C. B. and Roonwal, M. L. (1946). Studies on *Schistocerca gregaria* (Forskal). XIII. Food and feeding habits of the desert locust. *Indian J. Ent.* **8**, 141–163.

Large, R. V. (1973). Factors affecting the efficiency of protein production by populations of animals. *In* "The Biological Efficiency of Protein Production", (Ed. J. G. W. Jones), Cambridge University Press, London.

Roberts, N. F. (1954). Errors in yield appraisal in buying wool. *Wool Technol.* **1**, 77–83.

Ryder, M. L. and Stephenson, S. K. (1968). "Wool Growth", Academic Press, London and New York.

Shyamala, M. B., Sharada, K., Bhat, M. G. and Bhat, J. V. (1960). Chloromycetin in the nutrition of the silkworm *Bombyx mori* (L.). III. Influence on digestion and utilization of protein, fat and minerals. *J. Insect Physiol.* **4**, 229–234.

Soo Hoo, C. F. and Fraenkel, G. (1966). The consumption, digestion and utilization of food plants by a polyphagous insect, *Prodenia eridania* (Cramer). *J. Insect Physiol.* **12**, 711–730.

Yokoyama, T. (1963). Sericulture. *Ann. Rev. Entomol.* **8**, 287–306.

Waldbauer, G. P. (1968). The consumption and utilization of food by insects. *In* "Advances in Insect Physiology", Vol. 5, Academic Press, London and New York.

26

ANIMALS FOR WORK

Animals of many kinds have been employed for a very long time in a great variety of ways (Table 26.1) that have nothing to do with producing a product or, indeed, any material output.

The pollinating activities of bees have already been mentioned and the use of animals for transport, traction and even battle is well known. Less obvious is the use of animals in hunting and even fishing (e.g. the use of cormorants in the Far East).

Animals have been used as guards and alarm-raisers (from dogs to geese) and as companions. If the latter seems a rather trivial category, it is hard to draw useful and logical lines between, for example, guide-dogs for the blind, gun-dogs for the sportsman, watch-dogs for old people and pets for children.

In all these cases, the efficiency with which the animal does its job can be assessed in relation to its cost or to the resources used in keeping it. Measurement of performance may often be difficult but not necessarily more so than for alternative systems, such as burglar alarms instead of watch-dogs.

However, the agricultural use of animals for these sorts of purpose is somewhat less complicated, although still very varied (see Table 26.2) and with some overlaps with non-agricultural use.

Not all the animals used would normally be classed as "agricultural" but there is no good reason to exclude the sheepdog, for example, whilst including the buffalo used for cultivation.

The use of a dog to round up sheep and cattle is a good example, rather like insect pollination of crops and honey production by bees, of a function that it is difficult to imagine being carried out in any other way. It is true that people on horseback can manage without dogs, especially for herding cattle, and that the same task can be carried out in vehicles (even on tractors or motorcycles in some circumstances), but there are conditions in which there are no obvious alternatives at present. Of course, alternatives can always be

TABLE 26.1   Use of animals by man for purposes other than producing a product

Animal	Use
Horse, Camel	Traction, transport, recreation
Donkey (Ass), Mule, Oxen	Traction, transport
Llama	Transport
Dog	Traction, transport, herding, guarding, hunting, detection of drugs or explosives, tracking, guiding, companion, recreation
Cat	Hunting, companion
Elephant	Traction, transport
Monkey	Harvesting coconuts
Reindeer	Traction, transport
Ferret	Hunting
Rabbit, mouse, etc.	Laboratory animals (testing drugs, etc.)
Yak	Transport
Bees	Pollination
Insects	Biological control of weeds or other insects
Cormorants	Fishing
Pigeons	Carrying messages, recreation
Hawks, falcons, etc.	Hunting, recreation

Sources include: McKenzie (1974); Turner (1971)

TABLE 26.2   Agricultural use of animals for work by man

Work	Animals used
Traction, cultivation and clearing of land	Horses, oxen, buffalo, elephant
Transport	Horses, oxen, buffalo, donkeys, llamas
Pollination	Bees
Biological control of insect pests and weeds	Insects, arachnids, ducks, geese, amphibians, fish
Herding	Dogs

developed and animals can be trained instead of driven, attracted by feed or guided by well-placed fences.

The use of animals to perform these functions is always likely to involve less use of support energy, though not necessarily less labour, but will always carry a feed cost. This may be met from cheap by-products (e.g. offals for dogs) or it may involve substantial areas of land in producing forage. This major effect on land use is illustrated by the use of horses instead of tractors. It has been calculated (Blaxter, 1974) that the cultivation of a hundred hectares by horses, at current levels of UK power input, instead of by tractors, would require 30 hectares to be diverted to the production of feed for the working horses. (This assumes a need for one horse per two hectares of arable land.) In general, however, animal traction requires some 10–20% of the land surface to grow the feed for draught animals (de Wit, 1975). This, of course, greatly reduces the efficiency with which land, water, labour and fertilizer are used for the production of human food (and other agricultural products), but greatly increases the efficiency with which support energy is used. None of this is surprising and it emerges repeatedly that increases in the efficiency with which one resource is used may well result in, or be accompanied by, decreases in the efficiency with which other resources are used.

The relevance of efficiency calculations in relation to work animals appears greatest for transport and traction and least for functions such as pollination and, for example, biological control. In the case of pollination, the costs may be negligible, partly because the insects obtain their own feed from sources which would not be used at all otherwise.

Although pollination is normally dependent upon natural agencies, honey bees (*Apis mellifera*) are kept for this purpose in some numbers and two other species have also been "domesticated". These are alkali bees (*Nomia melanderi*) and leaf-cutter bees (*Megachile rotundata*), which have both been provided with artificial nest sites to encourage them amongst particular crops (alfalfa especially) in the USA (Pedersen and Garrison, 1973; Hanson and Barnes, 1973). Other wild bees are also cultured to some extent, including *Osmia* spp. (nesting in bamboo canes), humble bees (*Bombus* spp., nesting in boxes) and carpenter bees (*Xylocopa* spp., nesting in pre-pared softwood logs), and transported to the crops to be pollinated.

In the case of alkali bees, beds of earth are specially prepared, on a substantial scale (McGregor, 1976), and it has been calculated that a 1500 ft² (140 m²) bed would take care of 40 acres (16 ha). Bees will cover an area within a two-mile (three-kilometre) radius of their nest.

Some other social bees are used, notably the "stingless bees" (*Melipona* spp. and *Trigona* spp.), especially in Mexico.

Shemelkov (1960) calculated that one colony of honey bees was as effective as 300 men in the pollination of greenhouse cucumbers.

The importance of honey bees and cultured wild bees as pollinators increases as modern methods of agriculture result in greater areas of monoculture and decreasing numbers of wild pollinating insects. The most obvious advantage of adequate pollination of a crop is in increased yield but there are other benefits such as an earlier, more uniform crop in field bean (*Vicia faba*) or increased quality of the fruit, e.g. melon (*Cucumis melo*) and strawberry (*Fragaria × ananassa*) (Free, 1970). It has been estimated that the value of honey bees as pollinators is several times that of their value as producers of honey and wax.

Biological control by animals, whether control of weeds or of other animals, also involves no feed costs but other costs may not be negligible. The most important criterion is really how well the method works, however. Research costs may be high but cannot simply be allocated between the various biological control projects. There have been some spectacularly successful applications (Table 26.3) but far more important is the extent to which agriculture depends upon natural biological control: this is not itself operated by Man but may be greatly influenced by what he does. Where this method is deliberately employed, there may be substantial costs in rearing insects (or other organisms), in treating them (such as the exposure to radiation involved in the sterile-male technique (Bushland, 1971)), in distributing them and in monitoring the status of all the relevant populations.

As with chemical methods, adaptation of the organisms to be controlled can rapidly reduce the effectiveness of a particular method and there are always possibilities of unforeseen, undesirable consequences.

In many cases of this kind, the problems (of disease, loss and wastage)

TABLE 26.3　　Examples of successful biological control

Plant or insect to be controlled	Controlling organism
Prickly pear cactus in Australia (*Opuntia* spp.)	*Cactoblastis cactorum*
St John's-Wort in California (*Hypericum perforatum*)	*Chrysolina gemellata*
Insect pests in rice	Ducks
Greenhouse whitefly (*Trialeurodes vaporariorum*)	*Encarsia formosa*

Sources: DeBach (1964); Delucchi (1976); King (1966); DeBach (1974)

may be so great that ways have to be found to reduce their magnitude. Clearly, this needs to be done with the minimum of damage and risk and the cost has to be considered. It is probable that many different methods, including both biological and non-biological, will need to be used in an integrated manner, so the relative efficiency of any one method may be largely irrelevant. The potential for original thinking resulting in practical applications of great importance, seems to be enormous in this area.

Within those areas where efficiency ratios may be more useful, not a great deal of information is available. What there is will be considered under the headings of Transport and Traction.

Transport

The variety of burdens transported by animals defies description, varying from people to goods, other animals, raw materials and finished products. Furthermore, the object of transportation also varies greatly: sometimes speed is the crucial factor, sometimes the weight or volume moved, sometimes the distance over which it is moved, sometimes it relates to the terrain over which transport has to be accomplished, and sometimes to the climatic and weather conditions that have to be tolerated.

It is not possible, therefore, to describe one animal as more efficient than another for transport purposes, except for specified circumstances covering all these points. Clearly, some animals can carry heavier loads than others, because they are bigger or stronger, and some animals move faster than others. An illustration of the attributes that are known to influence transport efficiency is given in Table 26.4.

Traction

Most work of this kind involves the pulling of vehicles or equipment: in the former case, it overlaps greatly with transport. This is not always so, however, and some animal work involves pushing rather than pulling. Elephants are used to push trees over and pushing is part of their timber-moving technique. Working pumps and grinding machinery may also involve as much pushing as pulling and, in fact, it all depends upon the nature of the harness as to whether a working animal is chiefly doing the one or the other. It is obvious in any case that the efficiency with which a harnessed animal can work depends greatly on harness design, particularly in terms of points of attachment, and it is especially important not to confuse the problems of harnessing for work and those of controlling the animal when harnessed.

TABLE 26.5　Power of draught animals over long periods: (ploughing and tillage work)

Type and number of animals	Weight (kg)	Age (years)	Average effort (kN)	Mean of maximum efforts[a] (kN)	Speed (km h^{-1})	Power (kg m s^{-1})	Daily working hours		Duration of trial[b] (days)
							Hours worked	Effective hours	
1 donkey	160	—	0.450	0.863	—	—	6	3–3.5	14–10
1 pair of oxen N'Dama (Sefa)	657	6,8	0.883	1.667	2.2	54	5.5	5.5c	4–4
1 Pair of oxen N'Dama (Mirankro)	800	9,10	0.785	2.109	2.0	44	5	4	11–10
1 pair of Madagascar Zebu bullocks[d] (Kianjasoa)	650	4,5	0.785	1.471	2.5	56	4.3	4.3c	3–3

2 pairs of Madagascar Zebu bullocks	1 300	4,5	1.569	3.923	1.8	80	5.3	5.3[c]	2–2
3 pairs of Madagascar Zebu bullocks	1 945	4,5	1.961	4.266	1.6	88	5	5[c]	2–2
1 pair of oxen 1/2 Brahma (Miadana)	1 060	6	1.442	3.040	2.4	97	5.1	4.4	11–10

[a] The values shown in this colum are the mean of the maximum efforts recorded after each furrow

[b] The first figure indicates the period covered by the trial and the second the number of days worked

[c] The effective hours worked are identical with the hours worked since the animals were used morning and evening for 2 to 3 hours; it was not necessary to break these half-sessions with rest periods

[d] 3 pairs of Madagascar Zebu bullocks worked: first with 2 pairs in harness, then 3 pairs made up by adding a third pair, ploughing an unprepared field; subsequently a single pair of the original unit did harrowing

Source: F.A.O. (1972). The figures for effort have been converted from kilogram force (kgf) to kilo Newton (kN)

TABLE 26.4   Attributes of animals used for transport which influence their efficiency

Animal	Attributes
Buffalo	Docility, sure-footedness, strength
Camel	Feet adapted to walk on sand, ability to withstand heat and lack of water
Dog (Husky)	Coat adapted to withstand extreme cold, able to be trained to work in a team
Donkey	Strength and endurance for small size
Horse	Strength and speed
Mule	Endurance
Reindeer	Speed and adaptation to cold climates
Yak	Adaptation to high altitude and difficult terrain
Llama	Endurance at high altitudes

Sources include: Cockrill (1974); McKenzie (1974); Turner (1971)

The efficiency with which the job can be done also depends upon the design of the equipment or machinery and on the medium in which the animal has to work.

Buffaloes are adapted to cultivations in wet rice fields, where horses would be useless. Elephants are able to handle timber even in quite deep water and strong currents. Mules are generally used to carry things, but work in certain terrains requires their particular brand of sure-footedness.

Quite light animals (e.g. llamas) can pull carts and other wheeled vehicles (as can ostriches) and dogs can pull sleds; cultivation usually requires heavy animal (horses, cattle, buffalo), although this also depends on the nature of the soil.

The data available on the efficiency of animals for traction have been assembled in Tables 26.5 and 26.6.

TABLE 26.6   The efficiency of animals for traction

Animal	Av. wt. (kg)	Approx draught (kN)	Av. speed of work ms^{-1}	Power developed (W)
Light horses	400–700	0.588–0.785	1.0	735
Bullocks	500–900	0.588–0.785	0.6–0.85	550
Buffaloes	400–900	0.490–0.785	0.8–0.9	540
Cows	400–600	0.490–0.588	0.7	345
Mules	350–500	0.490–0.588	0.9–1.0	510
Donkeys	200–300	0.294–0.392	0.7	245

Source: F.A.O. (1969) The figures for draught have been converted from kilogram force (kgf) to kilo Newtons (kN)

Newton = The derived SI unit of force. The force required to give a mass of one kilogram an acceleration of one metre per second per second.

References

Blaxter, K. L. (1974). Power and agricultural revolution. *New Scientist* **61**, 400–403.

Bushland, R. C. (1971). Sterility principle for insect control: historical development and recent innovations (IAEA-SM-138/47). *In* "Sterility Principle for Insect Control or Eradication", Proc. of a Symp. Athens 1970. I.A.E.A., Vienna.

Cockrill, W. R. (1974). "The Husbandry and Health of the Domestic Buffalo", F.A.O., Rome.

DeBach, P. (1964). "Biological Control of Insect Pests and Weeds", Chapman and Hall, London.

DeBach, P. (1974). "Biological Control by Natural Enemies", Cambridge University Press, London.

Delucchi, V. L. (1976). "Studies in Biological Control", I.B.P.9. Cambridge University Press, Cambridge.

F.A.O. (1969). Farm implements for arid and tropical regions, by H. J. Hopfen. Rev. edn, Rome, F.A.O. F.A.O. Agricultural Development Paper No. 91.

F.A.O. (1972). Manual on the employment of draught animals in agriculture. F.A.O., Rome.

Free, J. B. (1970). "Insect Pollination of Crops", Academic Press London and New York.

Hanson, C. H. and Barnes, D. K. (1973). Alfalfa. *In* "Forage Crops", 3rd edn, (Eds M. E. Heath, D. S. Metcalfe and R. F. Barnes), Iowa State University Press, Ames, Iowa.

King, L. J. (1966). "Weeds of the World", Leonard Hill, London.

McGregor, S. E. (1976). Insect pollination of cultivated crop plants. Agricultural Handbook No. 496, A.R.S., U.S.D.A. Washington, D.C.

McKenzie, F. (1974). Contribution of working animals for power, herding, protection and pleasure. *In* "Animal Agriculture", (Eds H. H. Cole and Magnar Ronning), W. H. Freeman, San Francisco.

Pedersen, M. W. and Garrison, C. S. (1973). Legume and grass seed production. *In* "Forage Crops", 3rd edn, (Eds M. E. Heath, D. S. Metcalfe and R. F. Barnes), Iowa State University Press, Ames, Iowa.

Shemelkov, M. F. (1960). Particularities as to the utilization of bees for the pollination purposes of cucumber cultures in greenhouses and hotbeds. *In* "Nauchno-Issled. Inst. Ovoshchnogo Khoz", 49–58 (In Russian).

Turner, H. N. (1971). Conservation of genetic resources in domestic animals. *Outlook on Agriculture* 6(6), 254–260.

Wit, C. T. de (1975). Agriculture's uncertain claim on world energy resources. *Span* 18, 2–4.

27

THE USE OF FAECES

The quantities of organic wastes available each year, on a world basis, are enormous (Table 27.1), and a high proportion occurs as animal faeces, chiefly from the larger ruminants. In the UK, for example, cattle faeces amount to some 16.7×10^6 t DM per year, out of a total, including forestry, sewage sludge and refuse organic material, of about 36×10^6 t DM per year.

Faeces thus constitute a major resource. Historically, such material has been used primarily as a fertilizer or as a fuel, although in some parts of the world it has also been an important building material. More recent developments have included its use as a feed for animals, from conversion by invertebrates to recycling through the same species that produced it.

Fertilizer

Faeces and urine obviously contain most of the minerals consumed by animals, for only a small proportion is retained in the body. Composition varies with species (Table 27.2) and, since diet and digestive efficiency also vary, the daily output is much greater for some than for others (Table 27.3).

Provided that faeces and urine are returned to the land, crops used by animals should only require enough additional fertilizer inputs to restore what is removed in the animals themselves and to balance losses of various kinds. Losses occur when animals are housed or moved on roadways: slurry collected from houses may suffer losses during storage, handling, processing and distribution. On the other hand, excreta from housed animals can be spread more uniformly. Grazing animals distribute both faeces and urine in a very patchy manner and this may mean that some plants are oversupplied whilst others are deficient in nutrients.

The efficiency with which excreta is used as a fertilizer depends upon the extent of the losses, including gaseous ones, and the costs of all the processes

341

TABLE 27.1   Quantities of organic wastes produced in the world each year (not all of these quantities can be regarded as available currently)

Plant wastes	Estimated world production (tonnes × 10^6 DM)	
Straw		
wheat	295.9	
rye	26.4	
barley	120.3	
oats	48.2	
maize	473.1	
millet and sorghum	132.8	
rice, paddy	261.2	
Total		1 357.9
Sugar cane wastes		
tops	44.0	
bagasse	58.8	
Sugar beet tops	18.8	
Total sugar cane and beet wastes		121.6
Animal wastes		
excreta		1 796.0
Human wastes		
excreta		158.0
Total plant and animal wastes		3 433.5

Source: Duckham *et al.* (1976)

involved in making use of it. Advantages are often claimed for natural, as distinct from "artificial" fertilizers, in terms of effects on soil fauna and soil structure, over and above the minerals supplied. There undoubtedly are differences but they are not yet well quantified, in spite of the enormous period of time over which such fertilizers have been used.

Of course, animals are rarely, if ever, kept simply to produce faeces, although it has been pointed out that ruminants are quite efficient at it. In some cases, faeces may be one of the few usable products, where productivity is low or where religion or custom prevent the killing of cattle.

In most circumstances, however, faces and urine are by-products or, with

TABLE 27.2 Percentage composition of the excreta of various animals (average values after M.A.F.F., 1976)

	Moisture	N	P₂O₅	K₂O	Mg	Ca	Energy (MJ per kg DM)[a]
Poultry							
Battery hens — fresh droppings	75	1.5	1.1	0.6	0.12	0.8	14.6(1)
Turkeys — fresh droppings	55	1.8	1.4	0.9	—	—	—
Cattle							
Dairy cows — faeces	85	0.4	0.2	0.1	0.06	0.2	19.7–21.8(2)
Beef cattle — faeces	86	0.3	0.2	0.1	0.05	—	19.7–21.8(2)
Cattle — urine	92	0.79	0.002	1.6	0.02	—	—
Pigs							
faeces	76	0.6	0.8	0.4	—	0.1	
urine	97	0.4	0.1	0.5	0.002	—	
Sheep							
faeces	68	0.4	0.45	0.35	—	—	19.7–21.8(2)

[a]Sources: (1) Lowman and Knight (1970); (2) Barnes (1978)

TABLE 27.3 Daily output of excreta[a] by various animals (after Gowan, 1972)

	Faeces (kg per head)	Urine (kg per head)	Total excreta (faeces + urine) (kg per head)
Dairy cow	30	15	45
Beef animal	15	5	20
Sow	5	7	12
Fat pig	2	3	5
Sheep	2	0.5	2.5
Hen			0.09

[a]Excluding added water or litter

large concentrations of stock (such as are found in American beef feedlots), they may represent an unwanted output and thus a disposal problem, often with both monetary and environmental costs.

In either event, the most important implication is to the total efficiency of the animal production system in which the excreta occur. Whether it has a value or a cost, it can significantly affect overall efficiency in economic terms, or in terms of nitrogen utilization.

Animal Feed

Faeces represent undigested or partially digested feed, with some metabolic products and the remains of gut fauna and flora, all of which originally derived from the feed. They are thus, mostly, that part of the feed that is not incorporated in the product (or in the animal body that includes it) or used in its production. Clearly this is not quite true of all the faecal components and some waste is associated with all biological processes: furthermore, it ignores the importance of faecal matter as a carrier of metabolic wastes to be excreted and as a physical necessity for the proper functioning of the alimentary tract.

However, faeces are capable of further use by animals to produce more products, either more of the same or a totally different kind.

Faeces have been incorporated into the rations of cattle (Ward and Seckler, 1975; Bhattacharya and Taylor, 1975), chiefly, generally after drying (see Table 27.4). The support energy cost of drying is high (Table 27.5) and efficiency of support energy usage in animal production is generally decreased by feeding dried faeces. Incorporation in the wet form, on the other hand, would increase the risk of disease and raise problems of palatability, handling and distribution.

Of course, the recycling of dried faeces does have the effect of increasing the overall efficiency of feed use and thus of land use (Table 27.6). The use of wet faeces for feeding purposes has mainly been with fish, directly or indirectly (Eppel, 1963; Bardach *et al.*, 1972) and there are few quantitative data available relating to feed conversion rates. Aquatic systems may only be able to cope with limited quantities, however, unless additional oxygen is supplied and, even then, such systems may not be able to operate at low temperatures.

TABLE 27.4   Examples of the use of faeces in livestock feeding

| Species | Proportions of different faeces which have been satisfactorily incorporated into diets | | |
	DPW[a] (%)	Broiler litter (%)	Dried cow manure (%)
Broiler chicks	5	10	5–10
Laying hens	10		5–10
Growing pigs (over 32 kg)	10		
Beef (fattening)	55	25	20–27
	21		
	25		
	17.5		
Lambs (6 months old)	36		
Milking ewes	50	25	
		50	
Milking cows	30	75 of concentrate ration	

[a]DPW: dried poultry waste from caged layers

Source: Bhattacharya and Taylor (1975)

Other possibilities exist for using undiluted faeces but temperature is always likely to be a limiting factor in temperate countries. Two main avenues have been explored: the feeding of earthworms and of fly larvae.

TABLE 27.5   The support energy cost (fuel only) of drying poultry excreta

Using an example of a flock of 40 000 laying hens

Output   40 tonnes excreta per week at 75% moisture
         = 11 tonnes DPW[a] per week at 10% moisture
         = 10 tonnes DPW per week at   0% moisture

Input   4 546 litre oil
         or 414 l tonne^{-1} DPW

Gross energy content of DPW = 14.6 MJ kg^{-1} or 14 600 MJ tonne^{-1}
Energy content of fuel oil = 49 MJ l^{-1} or 20 286 MJ per 414 litre

Drying cost: 1.38 MJ per MJ of DPW

[a]DPW: dried poultry waste from caged layers

Sources: Blair (1975); Lowman and Knight (1970); Chapman *et al.* (1974)

TABLE 27.6   Effect of recycling faeces on efficiency of land use

No. of hens fed per hectare[a]	= 50
Annual production of faeces	= 1 643 kg
	= 411 kg DM
containing	118 kg crude protein
and	6 000 MJ gross energy
Annual CP requirement	˙ = 7 kg per laying hen
Additional no. of hens which could be supplied with CP requirement	= 17
Area of land which this would save	= 0.17 ha

[a]Assuming a ration of maize and soyabean meal and average European yields of these crops

Earthworms

Many species of earthworm have been cultured (Table 27.7) to produce bait for angling, to sell to farmers for introducing to soil, with the aim of improving soil fertility, or to feed to other animals, directly or after processing.

TABLE 27.7   Species of worm which have been cultured

Species	Common name	Use[a]
Lumbricus terrestris		Biodegradation of animal waste(1), food for pets
Helodrilus foetidus	Brandling manure worm	Fishing bait/biodegradation of rabbit manure(2)
Lumbricus rubellus	Red worm/Marsh worm	Fishing bait/biodegradation of rabbit manure(2)
Eisenia foetida	Brandling worm	Feed for broiler chicks and weanling pigs(3)
Enchytraeidae	White worms	Food for tropical fish(4)

[a]Sources: (1) Fosgate and Babb (1972); (2) Sabral (1978); (3) Sabine (1978); (4) Oughton (1970)

Such earthworms do not have to be reared on faeces but this does make a generally satisfactory, relatively cheap and plentiful medium.

The effect of culturing earthworms in faeces is to reduce the water and nitrogen content and these two processes are really the basis of the enterprise. The residual material has a higher value as potting soil, for gardeners and horticulturalists, than the original faeces, partly because of the lower water content. This, incidentally, represents one way in which material can be dried without the use of support energy.

The earthworms themselves are a means of recovering some of the nitrogen in the faeces and converting it into a protein feed (Table 27.8) that is suitable for feeding to pigs or poultry or as a constituent of pet foods.

The main disadvantages of earthworm culture are (a) that it is not a particularly pleasant process, (b) that it may take a considerable time (up to a year before harvesting) and (c) recovery of earthworms is laborious and messy. Of course, these features vary with the species, some of which breed

TABLE 27.8   Nitrogen recovered by earthworms from faeces

	Amount produced (kg per day)
Faeces from 1 000 cows	31 740
Conversion rate of raw faeces to live earthworms 10:1 ∴production of earthworms	3 174
Average DM content of earthworms is 22.9% ∴production of DM	727
Protein content of DM is 58.2% ∴production of protein	423
∴Protein produced	154 tonnes per year

Source: Fosgate and Babb (1972)

faster than others, and with the medium. Rabbit faeces seem particularly suited to the process but are not available in anything like the quantities that occur with cattle faeces.

It may also be that less laborious techniques of harvesting can be devised. Low temperatures might be used, for example, to cause the worms to leave the faecal beds.

Readiness to separate themselves from the culture medium is one of the advantages of fly larvae.

Fly larvae

Dipterous larvae (of both *Musca domestica* and *M. stabulans*) have been found to reduce poultry manure to a stable product in five to six days (Miller and Shaw, 1969; Calvert *et al.*, 1969). This process resulted in a loss of 80% of the organic matter and 30% of the moisture from the original material, at a temperature of 23–26 °C.

Details of the process are given in Table 27.9 and the efficiencies of energy and protein recovery are shown in Table 27.10. The resulting pupae have been fed to poultry, replacing soyabean meal and supplying 48% of the total protein (Calvert *et al.*, 1969).

One interesting feature of the use of fly larvae to convert poultry faeces is

TABLE 27.9   Reduction of poultry manure to a stable product by dipterous larvae

---

1. Newly laid house fly eggs seeded at the rate of three eggs per g of fresh poultry excreta

2. Kept at air temperature of 20–30°C with continuous overhead light for seven to eight days

3. Eggs hatch, larvae develop and by sixth day most have pupated

4. At the suggested seeding rate wt loss of excreta is in the region of 56% and the N content of the excreta is 2.04% compared with 5.6% in the fresh material

---

Source: Calvert *et al.* (1970)

TABLE 27.10   Efficiencies of energy and protein recovery from poultry manure by dipterous larvae

	Wt prod. (kg)	Moisture (%)	Protein content (kg)	Energy content (MJ)
Input				
Fresh droppings from 1 000 battery hens per day	100	75.0	8.75	365
Output				
Pupae	2.28	66.8	0.5	15.9
Larvae	0.05	72.8	0.1	2.7
Residual compost	44.3	50.0	2.8	—

Calculated from: Calvert *et al.* (1970)

that there is an optimum "seeding" rate of about three fly eggs per gram of wet faeces (Calvert *et al.*, 1970). This is because too few larvae produce less than the maximum output per gram of faeces, whereas too many larvae result in rather undersized individuals (Fig. 27.1). This bears many similarities to the situation with grazing animals (see Chapter 18), except that fly larvae pupate after a given interval, rather than at a given size. This means that overcrowded larvae produce smaller pupae, whereas smaller grazing animals can be kept longer in order to gain the same amount of weight as less heavily stocked animals.

However, it would be expected that for cold-blooded fly larvae, the same total production would have been obtained for all populations, once an optimum "stocking density" had been passed. This does not mean, of course, that the individual size of the pupa is unimportant.

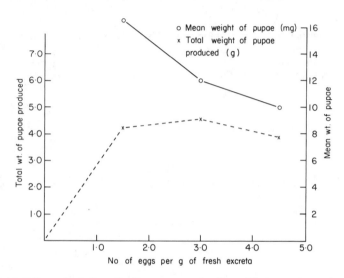

FIG. 27.1   Effect of egg "seeding" rate on production of fly larvae in poultry faeces (after Calvert et al., 1970).

References

Bardach, J. E., Ryther, J. H. and McLarney, W. O. (1972). "Aquaculture", Wiley-Interscience, New York.

Barnes, R. J. (1978). Personal communication. Grassland Research Institute, Hurley, Berkshire.

Bhattacharya, A. N. and Taylor, J. C. (1975). Recycling animal waste as a feedstuff: a review. *J. Anim. Sci.* 41(5), 1438–1457.

Blair, R. (1975). Poultry Research Centre. Personal communication.

Calvert, C. C., Martin, R. D. and Morgan, N. O. (1969). House fly pupae as food for poultry. *J. Econ. Ent.* 62(1), 938–939.

Calvert, C. C., Morgan, N. O. and Martin, R. D. (1970). House fly larvae: bio-degradation of hen excreta to useful products. *Poultry Sci.* 49 I, 588–589.

Chapman, P. F., Leach, G. and Slesser, M. (1974). 2. The energy costs of fuels. *Energy Policy* 2(3), 231–243.

Duckham, A. N., Jones, J. G. W. and Roberts, E. H. (1976). Food Production and Consumption. North Holland Publishing Co., Amsterdam.

Eppel, S. (1963). Development and practice of duck fattening on natural waters in conjunction with carp breeding. *Arch. Geflugelzucht Kleintierk* **12**, 30–34.

Fosgate, O. T. and Babb, M. R. (1972). Biodegradation of animal waste by *Lumbricus terrestris*. *J. Dairy Sci.* **55(6)**, 870–872.

Gowan, D. (1972). "Slurry and Farm Waste Disposal", Farming Press, Suffolk.

Lowman, B. G. and Knight, D. W. (1970). A note on the apparent digestibility of energy and protein in dried poultry excreta. *Anim. Prod.* **12**, 525–528.

M.A.F.F. (1976). Organic manures. Bulletin 210. H.M.S.O., London.

Miller, B. F. and Shaw, J. H. (1969). Digestion of poultry manure by diptera. *Poult. Sci.* (abstr.) **48 II, (5)**, 1844.

Oughton, J. G. (1970). Fishworms. Agdex, 610. Ontario Department of Agriculture and Food.

Sabine, J. R. (1978). The nutritive value of earthworm meal. *In* "Utilization of Soil Organisms in Sludge Management", 122–130. National Tech. Information Services, Springfield, U.S.A.

Sabral, R. (1978). "Earthworm Farming", Salop Rabbit and Worm Producers Ltd, Whitchurch, Salop.

Ward, G. M. and Seckler, D. (1975). Recycling the protein of animal waste: protein to support animal protein production. *World Rev. Anim. Prod.* **11(1)**, 54–59.

Part IV
The Relative Efficiency of Plants and Animals

28

THE RELATIVE EFFICIENCY
OF PLANTS AND ANIMALS

As was pointed out in Chapter 1, the efficiency of different production processes can only be compared for some common purpose and such comparisons usually have to be based on the use made of the same (or a very similar) resources and the output of sufficiently similar products. A potato crop has no efficiency in the production of tomatoes and its use of, say, water cannot be compared with the water use of some other crop except in terms of a common product (potatoes, dry matter, energy, protein and so on).

The relative efficiency of crop plants and agricultural animals can always be related to similar resources, because animal production must ultimately be based on plant production. The efficiency with which grasses are used by grazing animals cannot generally be compared with anything that plants do (since few plants live by eating other plants), except by expressing the resource used in, for example, energy terms.

Looked at in this way, however, animal production nearly always appears much less efficient than crop production, unless a higher weighting is given to the output of animal production. This can most readily be done in monetary terms.

Thus energy and protein production are always much higher from crops than from animals, on land that could be used for either crop or animal production, whether expressed per hectare or per unit of solar radiation received (Tables 28.1 and 28.2). But the output values look rather different when expressed in monetary terms (Table 28.3). This reflects relative supply and demand, of course, but also the fact that most of us, where we have an effective choice, generally place a higher monetary value on a unit of dietary energy or protein when it is of animal origin and especially, in particular, well-recognized and well-liked forms, such as beef steak.

The most extreme views of the relative efficiencies of animal and crop production are therefore to be found (a) where the satisfaction of the

355

TABLE 28.1(a) Production of crude protein and gross energy by crops (from Spedding, 1981)

Crop	Yield component	Yield (kg)	DM (%)	Yield DM (kg ha⁻¹)	Crude Protein (%)	Gross energy (MJ kg⁻¹)	Crude protein (kg ha⁻¹)	Gross energy (MJ ha⁻¹)	Sources[a]
Grass (perennial ryegrass)	Total harvested	60 000	20.0	12 000	17.5	18.5	2 100	222 000	(1) (2)
Cabbage	Total harvested	54 545	11.0	6 000	13.6	17.5	816	105 000	(3) (2)
Field beans (spring)	Seed	2 511	86.0	2 159	31.4	19.0	678	41 021	(4) (2)
(winter)	Seed	2 688	86.0	2 312	26.5	18.8	613	43 466	(4) (2)
Peas	Seed	2 511	86.0	2 159	26.2	18.9	566	40 805	(4) (2)
Potatoes (maincrop)	Tuber	27 621	21.0	5 800	9.0	17.6	522	102 080	(4) (2)
Wheat (winter)	Seed	4 394	86.0	3 779	12.4	18.4	469	69 534	(4) (2)
Sugar beet	Root	37 666	23.0	8 663	4.8	17.6	416	152 469	(4) (2)
Maize	Seed	4 654	86.0	3 995	9.8	19.0	392	75 905	(4) (2)
Rice	Seed	5 670	86.0	4 876	7.7	18.0	375	87 768	(5) (2)
Barley	Seed	3 767	86.0	3 239	10.8	18.3	350	59 274	(4) (2)

[a]Sources: (1) Spedding and Diekmahns (1972); (2) M.A.F.F. (1975); (3) Holmes (1971); (4) Nix (1974); (5) F.A.O. (1975).

TABLE 28.1(b) Production of crude protein and gross energy by animals

Animal	Yield component	Yield (kg ha⁻¹)	Crude protein (%)	Gross energy (MJ kg⁻¹)	Crude protein (kg per ha per yr)	Gross energy (MJ per ha per yr)	Sources or data used[a]
Cattle, beef (suckler)	Carcase	360	14.8	10.9	53	3 924	(1) (9)[a]
(barley)	Carcase	440	14.8	10.9	65	4 796	(2) (9)
Cattle, dairy	Milk	3 386	3.5	2.59	118.5	8 770	(3) (9)
Sheep	Carcase	462	14.0	16.203	65	7 486	(4) (9)
Pigs (baconer)	Carcase	875	11.95	16.5	105	14 438	(5) (9)
Rabbit	Carcase	1 511	19.31	8.77	292	13 251	(6)
Hen	Carcase	980	13.8	7.2	135	7 056	(7) (9)
Hen	Eggs (edible portion)	624	11.9	6.6	74	4 118	(8) (9)

[a]Sources: (1) J.B.P.C. (1972). All grassland system. Stocking Rate of 1.5 cows and calves per hectare, including allowance for concentrates. No allowances for sire, replacement or mortality. Energy and protein content of the carcase estimated from Refs in the literature. (2) J.B.P.C. (1968). Intensive beef system using calves from the dairy herd. No allowances for sire, dam, replacement or mortality. (3) Walsingham and Bather (1976). Stocking Rate of 0.8 cows per hectare — allowing for concentrates. No allowance for replacements. (4) Young and Newton (1975). No allowance for ram or replacements. (5) Data calculated from recommended diets for pigs and assuming sow producing 20 progeny per year; 1 boar to 20 sows. No allowance for replacements or mortality. (6) Data from J. M. Walsingham, based on a ration of 70% lucerne (yield of 11 000 kg DM per year) and 30% barley. Doe producing 64 progeny per year 1 buck to 20 does. No allowance for mortality or replacements. (7) Based on feed consumption of broiler and share of parents' feed; no allowance for mortality or replacements. (8) Feed consumption of hen during rearing and laying. (9) Concentrate rations based on barley with beans added where necessary to give the required protein levels.

TABLE 28.2   Production of protein and energy per unit of solar radiation received[a]

	Crude protein produced per MJ solar radiation received per annum (kg)	Energy produced per MJ solar radiation received per annum (MJ)
Crop		
Grass (perennial ryegrass)	0.00006	0.007
Cabbage	0.000025	0.003
Field beans		
(spring)	0.000021	0.0012
(winter)	0.000019	0.0013
Peas	0.000017	0.0012
Potato (maincrop)	0.000016	0.003
Wheat (winter)	0.000014	0.002
Sugar beet	0.000013	0.005
Maize	0.000012	0.002
Rice	0.000011	0.0026
Barley	0.000011	0.0018
Animal		
Cattle beef		
(suckler)	0.0000016	0.00012
(barley)	0.000002	0.00015
Cattle dairy	0.0000036	0.00027
Sheep	0.000002	0.00023
Pigs (baconer)	0.0000032	0.00044
Rabbit	0.0000088	0.00040
Hen	0.0000041	0.00021
Hen (eggs)	0.0000022	0.00012

[a]Production details as for Table 28.1, solar radiation received is 33×10^6 MJ per ha per year i.e. the sort of figure expected for the UK

wealthy is the criterion and (b) where the main concern is the feeding of the maximum number of people per unit of resource employed.

In the latter case, there will generally be a heavy emphasis on crop production, since, where crops can be grown, more people can be supported in this way (Table 28.4).

However, as has been pointed out in earlier chapters, there are large areas of land that will not support crop production at all or will not support it at a level that renders it either practicable or economic (though this depends greatly on the cost and availability of labour). On such areas, animal

TABLE 28.3   Production of protein and energy expressed in monetary terms

Product	Output (kg ha⁻¹)	Price[a] (p kg⁻¹)	Output (£ ha⁻¹)	Value[b] p/kg CP	p/MJ energy
Crop					
Grass (hay, 85% DM)	14 118	4.1	579	27.6	0.26
Cabbage	54 545	6.0	3 273	40.0	3.12
Field beans (winter)	2 688	5.6	151	24.6	0.35
Peas (marrowfat)	2 511	10.3	259	45.8	0.63
Potatoes (maincrop)	27 621	2.9	801	153.4	0.78
Wheat (winter)	4 394	5.7	250	53.3	0.36
Sugar beet	37 666	1.3	490	117.8	0.32
Maize	4 654	5.4	251	64.0	0.33
Barley	3 767	5.8	218	62.3	0.37
Animal					
Beef (barley)	440	60.0	264	406.2	5.5
Lamb	462	76.5	353	543.1	4.7
Pork	875	49.0	429	408.6	3.0
Chicken	980	46.0	451	334.1	6.4
Rabbit	1 511	80.0	1 209	414.0	9.1
Eggs	624	38.5	240	324.0	5.8
Milk	3 386	5.0	169	143.0	1.93

[a]All prices as paid to farmers 1974–5 (except for milk) from: M.A.F.F. (1977) Agricultural Statistics England and Wales 1975, H.M.S.O., London
[b]In each case the crude protein or energy component is bearing the whole cost
Milk price 1973–74 from M.M.B.

Other output data from Table 28.1.

production may represent the most efficient way of collecting, concentrating, and converting the natural vegetation to human food. For large concentrations of human beings able to use these areas as *additional* sources of food, this is certainly true. Whether it necessarily follows that, on that land alone, animal production would support the maximum number of people, is less certain.

Take, for example, a hill sheep farming situation on the uplands of the UK. The carcase output from sheep would amount to about 144 kg per hectare per year. (This assumes that the enterprise would be primarily a producer of meat: this is not likely in practice, where wool and breeding stock would be more probable outputs. Nevertheless, in considering how many people could be fed per unit of land, it has to be viewed in this way, in order to feed anyone at all.)

TABLE 28.4 The number of people who could be supported by the production from one hectare of land

| | Crude protein (kg ha^{-1}) | Gross energy (MJ ha^{-1}) | No. of people whose annual requirement could be met | |
			Protein[a]	Energy[b]
Crop				
Cabbage	816	105 000	34	23
Field beans	613	43 466	26	9
Peas	566	40 805	24	9
Potato	522	102 080	22	22
Wheat	469	69 534	20	15
Sugar beet	416	152 469	17	33
Maize	392	75 905	16	17
Rice	375	87 768	16	19
Barley	350	59 274	15	13
Animal				
Beef	65	4 796	3	1
Lamb	65	7 486	3	2
Bacon	105	14 438	4	3
Rabbit	292	13 251	12	3
Chicken	135	7 056	6	2
Eggs	74	4 118	3	1
Milk	118	8 770	5	2

[a]Taking protein requirement as 65 g per day i.e. 24 kg per year (assuming adequate protein quality)

[b]Taking energy requirement as 12.6 MJ per day i.e. 4599 MJ per year (assuming all the energy is available e.g. digestible)

This means that one person would require the produce of 3.8 hectares in order to receive an adequate diet (of *c.* 12.6 MJ and 65 g protein per day). (The area of 3.8 ha is needed to cover the energy requirements; it would result in a surplus of protein, assuming that protein quality was adequate. The quantities of energy and protein are recommended by UK Department of Health.) Now this could be obtained from 1256 kg of potatoes, requiring about 0.045 hectare, for example. (This assumes that the gross energy and protein contents are available to humans and that the potatoes are not peeled.) In other words, quite small areas with average yields or a somewhat greater area with very low yields could supply an equivalent quantity of energy and protein (Table 28.5). Furthermore, the labour required in total would be no greater (and probably much less) and it would be concentrated

TABLE 28.5   Land area required to supply protein and energy requirement for one person, from potatoes

Yield of potatoes(1)[a] (kg ha^{-1})	Area required to provide one person with energy and protein requirement(2) (ha)
20 000 low	0.06
28 000 average	0.04
36 000 high	0.03

[a](1) Nix (1976); (2) Taking 12.6 MJ energy and 65 g protein per day as the requirement of one person and the protein and energy content of potatoes as for Table 28.1

largely in one half of the year. If very high yields were sought, on a gardening basis, only 0.01 hectares would have to be cultivated intensively with high inputs of fertilizer and water.

In both cases, whether using animal or crop production, one person's efforts would have to yield more than simply the satisfaction of dietary needs but, in terms of feeding people actually living on the land, it is hard to see how animal production could be the primary choice. In practice, it is almost certain that a mixture of animal and crop production would be the best solution. The point to be made, however, is that the efficiency with which animals can be used depends not only on the nature of the land but also on the ways in which the human population is organized.

There are some areas of natural vegetation where it can be argued that controlled use of the indigenous animal species is the most efficient way of utilizing the total environment, since these species are adapted to the climate, terrain and diseases. This has often been claimed for parts of Africa (Harrison Matthews, 1963; Talbot et al., 1965; Ledger et al., 1967; UNESCO, 1979), although Tribe and Pratt (1975) concluded that "wildlife are unlikely in most situations to be more efficient than domestic animals in satisfying the needs of man". Nevertheless, based on the same general idea, there are possibilities for the control and domestication of hitherto unused species (see Table 28.6).

Whether crop production can be sustained at any particular level of yield depends almost entirely on the nature and level of the inputs that can be afforded. If land is in very short supply and food is urgently required, most of the difficulties of harvesting and cultivation can be overcome by a high input of human labour. The same may be true of pest and weed control and, to

TABLE 28.6   Wild species of animals being considered for domestication

Animal	Place	Reference
Red deer (*Cervus elaphus*)	Scotland	Bannerman and Blaxter (1969)
Eland (*Taurotragus oryx*)	Africa Russia	Kyle (1972); Krutyporoh and Treus (1969)
Capybara (*Hydrochoerus hidrochaeris*)	South America	Jimenez and Parra (1975)
Grass cutter rat (*Thryanomys swinderianus*)	West Africa	Hartog and Vos (1973)
Turtle (*Chelonia mydas*)	West Indies Australia	Ulrich and Parkes (1978); Bustard (1972)

some extent, fertilizer inputs can be assured by detailed attention to re-cycling (including all human wastes). Climatic limitations are more difficult to eliminate, although labour may be an important component of building costs and thus a degree of protection could be paid for in these terms. Furthermore, it can be argued that where climatic restrictions are severe for a large part of the year, maximizing crop production per unit of land and time during the available growing season is a better way of using the natural resources than by keeping animals (throughout the year) or by trying to grow crops outside the natural growing season.

Certainly it is true that countries such as Finland may produce quantities of herbage during their short growing season similar to those produced at lower latitudes in the UK.

Similar considerations apply to the utilization of land where the climate includes a severe dry season, although there are clearly possibilities for water (the scarce commodity) to be recycled in protected cropping systems (Morgan, 1979). Indeed, there is even the possibility that the recycling of transpired water could provide an additional source of energy (Morgan and Bather, 1980).

Clearly, few worthwhile generalizations can be made about the relative efficiency of crop and animal production systems. In many situations, the optimum agricultural system will contain both crop and animal production components and the problem will be to achieve the right balance. In these circumstances, the optimum human diet would be an omnivorous one but the animal part of it might not be what most people would actually prefer.

Decisions about preference are personal ones: decisions about whether such preferences can be afforded may be either personal or social (relating to a community, a nation or the population of the world).

There are clearly both economic and moral issues involved in this argument and it may be that all would be well if economic values adequately reflected moral values. The problem can be illustrated by one of the commonest versions of the argument: is it right to feed grain to animals?

The underlying assumptions are (a) that more people could be fed on the grain than on the resulting animal products (which is correct), (b) that the grain is not surplus to human needs and (c) that the animal products are not strictly necessary.

The difficulties, of course, are that "surplus to human needs" is not the same as "surplus to demand". Thus those who need the grain may not be able to pay for it and there may be no apparent way of putting this right. Whilst such a socio-political problem exists, simply diverting grain from animal feeding could have most unfortunate effects.

Furthermore, the grain may not be in the right place at the right time and considerable costs are involved in both transport and storage.

If all these difficulties were overcome and if the total amount of food produced from crops were only sufficient for the needs of the world population, then it would be irrational to feed crops to animals, if these crops could be consumed directly, since some people would obviously have less than enough to eat as a result.

Currently there is or could be a surplus but it has been calculated (Reid, 1970) that, given certain assumptions about the growth of population and need for food, the genuine surplus of crop production over requirement would diminish, so that by 1985 considerably less concentrates would be available for milk and meat production by ruminants and by 2000 no concentrates would be available for the production of eggs, poultry, milk and ruminant meat.

Of course, some would argue that efficiency of animal production can be greatly increased and that new species could be domesticated (Short, 1976; Short, 1977). Others have pointed to the enormous potential for animal feeding (Dyer et al., 1975) represented by the vast quantities of cellulosic waste, estimated on a world basis to be 100 000 million tons per annum (Microwaves, 1974). Based on this estimate, it has been calculated (Dyer et al., 1975) that the energy required to provide 100% of the 5.98×10^{10} kg world's annual protein need could be met by feeding only 5% of the world's waste cellulose to cattle, sheep and goats. Unfortunately, no estimate of the energy cost of doing this is available.

Similarly, orthodox farming, such as beef production from pasture in the tropics, could be enormously increased (Stobbs, 1975), although this might involve very large fertilizer inputs (though not necessarily of nitrogen).

There is, of course, no good reason why the world should try to feed the maximum possible population, since food is not the only factor of importance and there is every reason to stop population growth at a point where none of these limiting factors is operating. That is simply to say that a conscious decision is better than allowing any of the other limiting factors to operate. Everyone would probably agree with this, provided that satisfactory means could be found to make the decision effective.

In the meantime, world population will continue to grow for some time (Bunting, 1974) and there will be a need to increase food production to meet this increasing requirement. One often neglected facet of population growth is that the only resource to become more abundant in consequence is that of labour. Once it is accepted that more labour might be used in agriculture, the outlook is entirely changed, in terms of the appropriate *scale* of enterprise, methods of cultivation, weed and pest control and harvesting. Indeed, hand harvesting for a household's food needs points to quite different crops, crop mixtures and sequences. For machine harvesting, there are advantages in uniformity of product, time of ripening and so on. For hand harvesting and feeding the producer, uniformity may even be a disadvantage and as wide a spread of harvesting date as possible may be desired.

Perhaps the most important outcome of the discussion so far is the recognition that increasing *efficiency* of production is a worthwhile aim, independent of the total amount of food that is required. Even if we were producing enough in total, it would be better if the efficiency with which it was produced could be increased, since this would decrease its cost, in real terms (of resources used).

Of course, as should be evident by now, it is necessary to decide which aspects of efficiency should be the subject of improvement and this presents many problems.

In general, however, no progress is likely to be made until such a decision is arrived at, with whatever definition of constraints should prove to be necessary in relation to other aspects.

Once an efficiency ratio has been specified, it is possible to consider ways in which it could be improved.

The Improvement of Efficiency

Improvement cannot be deduced from the ratio itself simply because no interactions between numerator and denominator are stated. Thus the efficiency of meat (M) production from feed (F) can be expressed as M/F and could be improved by increasing M or decreasing F, but not by *all* such increases and decreases. For example, reducing or eliminating F would not necessarily improve the ratio, since it would almost certainly also reduce M, probably to a disproportionate extent. This is because M and F are not independent of each other.

It is quite impossible to explore ways of improving efficiency and to assess their effect, unless the ratio is expanded (see Fig. 28.1) to include (a) the main factors and mechanisms influencing both M and F and (b) the main interactions between these factors. This amounts to a simple modelling exercise, to describe the system to be improved in terms that are adequate for the purposes of identifying potential improvements and assessing their probable effects.

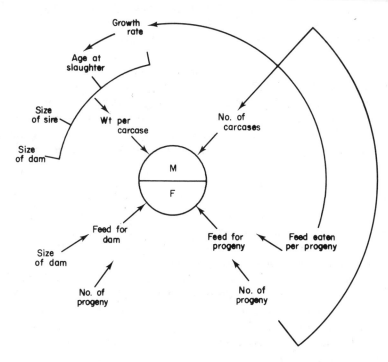

FIG. 28.1 Expansion of a ratio expressing efficiency of meat production.

It is hard to see how complex agricultural systems can be sufficiently understood for these purposes without some attempt to express all the components and their relationships in quantitative terms. Such mathematical models allow a great deal of quantified thinking to be based on an unambiguous statement of the relevant facts and sensitivity tests to be carried out in relation to changes in the most important variables. Predictions can then be made that are clearly based on stated assumptions.

Changing input values in real systems is often a very poor way to discover anything of value, because the situation is complex and the effects of many variables are confounded with those of others, including those which cannot be controlled (e.g. the weather). The more complex the real system, the more value there is in a theoretical exploration of the possibilities of improvement. Naturally, the resulting hypotheses have to be tested in real situations but the number of hypotheses worth examining in this way would be limited. A crucial question in a discussion of biological efficiency in agriculture concerns the extent to which important improvements in agricultural systems will come from increased efficiency of the biological processes within them. Furthermore, some of the biological processes might involve less familiar animals (such as insects and snails), unusual processes (such as the use of fly larvae to dry waste faeces (Miller and Shaw, 1969; Calvert *et al.*, 1970)), and the utilization of areas not currently farmed to any great extent. In the last category, there may be major developments in aquaculture (see Shepherd, 1975). Biological control (by insectivorous birds as well as insects?), highly protected cropping systems and unusually integrated crop and animal systems may be devised, that combine many different advantages. Energy crops, including tree plantations, could be grazed or used to produce poultry, and insect-eating poultry could be used to control pests in crops.

It is, of course, in the nature of scientific research that advances cannot necessarily be foreseen, so it is not possible to answer the question with any certainty. It *is* possible, however, to consider what aspects of efficiency are likely to be of the greatest importance in the future.

As far as products are concerned, it is only possible to make some very general statements. For example, it seems probable that agriculture will remain by far the most important source of human food and that food production will continue to be the major purpose of agricultural activities. It is not possible to say much about the form of this food, except that processing methods may make the initial form of less consequence (but this is affected by one's views about whether processing will make an increased or decreased contribution).

None of this necessarily implies that other agricultural products, such as raw materials for fuel, clothing and other industrial uses, will decrease in importance. Indeed, it seems quite possible that they would all become more important.

As soon as the reasons for this are examined, it becomes clear that *if* such outputs become of greater importance, it will not be primarily because of their inherent qualities but because the main (energy) resource used is virtually renewable.

Indeed, the strongest case for agriculture is that it is based on solar radiation. Although this costs something to collect and use, it is available in vast quantities and will be so for the foreseeable future. Furthermore, the amount available tomorrow is quite uninfluenced by the amounts used, or wasted, today. This is not true for any other resource, although many of them can be efficiently recycled (at some energy cost), and it is particularly inapplicable to support energy.

Unless a safe, abundant and relatively cheap source of energy becomes available (and some argue that this will be so, by some nuclear route), agriculture will have to depend primarily and increasingly on solar radiation. Now, as was pointed out in Chapter 1, this is the reverse of what has been happening in the developed agricultural systems of the world.

Furthermore, at present, only biological systems make significant use of solar radiation: it therefore seems probable that the biological content of agricultural systems may have to increase.

It is important to be clear about the reasons for this. The main aim must be to increase the use of solar radiation whilst decreasing the use of support energy. A major reason for the input of support energy to agriculture has been to *increase* the efficiency with which solar radiation is used. This is most obvious with fertilizers, where the efficiency of use of solar radiation may be increased by the input of fertilizer; it is also true for almost all resources except support energy. Certainly, more is produced per unit of land, labour, water, time and money, up to some optimal level of fertilizer application.

In general, where support energy inputs are used to increase output (and this has been estimated as about one third of the total used in developed agricultural systems (de Wit, 1975)), the efficiency of use of solar radiation will be increased.

However, as much support energy may be used in order to save labour as is used to increase production per unit of land. This represents some diversion of solar radiation (used to produce food for the labourer) and a substitution of support energy (though not necessarily proportional) without any increase in total output (per hectare).

The problem of decreasing the support energy inputs of all kinds, without reduction in food production, is not to be solved simply by trying to eliminate those inputs that do not directly affect production.

For example, legumes may be able to replace fertilizer nitrogen without reduction in yield per hectare, if not immediately, then by improvements in plant breeding or in the efficiency of nodulation or bacterial fixation of nitrogen, or by better pest control, and so on.

There is no simple way of determining where reductions in support energy input can best be made but there is a great need to know where the inputs occur and what role they play. What is then clear is that only biological processes can be used to effect a substitution, since the energy has to come from somewhere and this is the main way of utilizing solar radiation. It may not be the only way but it will certainly remain the most important and the least likely to require scarce material resources as well.

It needs to be recognized that biological processes do have their essential requirements, such as oxygen, some of which, such as water, may also be in short supply.

Even so, the single most important aspect of agricultural efficiency in the future is likely to be that of energy use and it is probable that the biological efficiency with which energy, solar and non-solar, is utilized for the total production of agricultural outputs will be the most important of the efficiency ratios with which we are concerned.

References

Bannerman, M. M. and Blaxter, K. L. (1969). The husbandry of red deer. Proc. Conf. Rowett Inst., 1969. Highlands and Islands Dev. Board and Rowett Research Inst.

Bunting, A. H. (1974). Population, agriculture and poverty in the developing world. Public Lecture, University of Reading, Jan. 1974.

Bustard, R. (1972). "Sea Turtles", Collins, London and Sydney.

Calvert, C. C., Morgan, N. O. and Martin, R. D. (1970). House fly larvae: biodegradation of hen excreta to useful products. *Poult. Sci.* **49** I, 588–589.

Dyer, I. A., Riquelme, E., Baribo, L. and Couch, B. Y. (1975). Waste cellulose as an energy source for animal protein production. *Wld Anim. Rev.* No. 15, 39–43.

F.A.O. (1975). "Production Yearbook, 1974", Vol. 28. F.A.O., Rome.

Harrison Matthews, L. (1963). A new development in the conservation of African animals. *In* "The Better Use of the World's Fauna for Food", (Ed. J. D. Ovington), 37–44. Institute of Biology, London.

Hartog, A. P. den and Vos, A. de (1973). The use of rodents as food in tropical Africa. F.A.O. *Nutrition Newsletter* **11(2)**.

Holmes, W. (1971). Efficiency of food production by the animal industries. *In* "Potential Crop Production", (Eds P. F. Wareing and J. P. Cooper), 213–227 Heineman, London.

Jimenez, E. Gonzales and Parra, R. (1975). The capybara, a meat producing animal for the flooded areas of the tropics. *In* "Proc. III World Conf. Anim. Prod", (Ed. R. L. Reid), Sydney University Press, Australia.

Joint Beef Production Committee (1968). Handbook No. 2. Intensive cereal beef system using calves from the dairy herd. M.L.C., Bletchley, Bucks.

Joint Beef Production Committee (1972). Handbook No. 3. Suckled calves. Meat and Livestock Commission, Bletchley, Bucks.

Krutyporoh, F. and Treus, V. (1969). Domestication of Eland. *Moloch. Myas. Skotov., Mosk.* **14 (8)**, 36–38 (In Russian).

Kyle, R. (1972). "Meat Production in Africa — The Case for New Domestic Species", University of Bristol, England.

Ledger, H. P., Sachs, R. and Smith, N. S. (1967). Wildlife and food production. *World Rev. Anim. Prod.* **111**, 13–37.

M.A.F.F. (1975). Tables of Feed Composition and Energy Allowances for Ruminants. H.M.S.O., London.

Microwaves (1974). New Brunswick Scientific Co.

Miller, B. F. and Shaw, J. H. (1969). Digestion of poultry manure by diptera. *Poult. Sci.* **48, II**, 1844 (abstr.).

Morgan, G. H. (1979). The effect of salinity on transpiration in *Beta vulgaris* a study with reference to the possible recovery of transpired water vapour in closed system greenhouses. Dissertation for B.Sc. (Hons), Biological Science, School of Environmental Sciences, Plymouth Polytechnic.

Morgan, K. E. and Bather, D. M. (1980). Energy — making the most of it. Paper presented at the 28th National Power Farming Conf., Blackpool, 1980.

Nix, J. (1974). "Farm Management Pocketbook", 6th edn Wye College, Kent.

Nix, J. (1976). "Farm Management Pocketbook", 7th edn Wye College, Kent.

Reid, J. T. (1970). Will meat, milk and egg production be possible in the future? Proc. 1970 Cornell Nutr. Conf., 50–63. Cornell University Press, Ithaca, N.Y.

Shepherd, C. J. (1975). The economics of aquaculture — a review. *Oceanogr. Mar. Biol. Ann. Rev.* **13**, 413–420.

Short, R. V. (1976). The introduction of new species of animals for the purpose of domestication. *Symp. Zool. Soc. Lond.* No. 40, 321–333.

Short, R. V. (1977). Wild animals — "an untapped source of information", Report of the Sir William Weipers Lecture. *Vet. Rec.* **100(21)**, 439–440.

Spedding, C. R. W. (1981). New horizons in animal production. *In* "Animal Production Systems", (Ed. B. L. Nestel), Vol. 2A of World Animal Science, Elsevier, Amsterdam (in press).

Spedding, C. R. W. and Diekmahns, E. C. (1972). "Crop Grasses and Legumes in British Agriculture", Commonwealth Agricultural Bureaux, Farnham Royal, Buckinghamshire.

Stobbs, T. H. (1975). Beef production from improved pastures in the tropics. *Wld Rev. Anim. Prod.* **XI(2)**, 58–65.

Talbot, L. M., Payne, W. J. A., Ledger, H. P., Verdcourt, L. D. and Talbot, M. H. (1965). The Meat Production Potential of Wild Animals in Africa. Tech. Comm. No. 16. Commonwealth Bureau of Animal Breeding and Genetics, Edinburgh. C.A.B., Farnham Royal, Buckinghamshire.

Tribe, D. E. and Pratt, D. J. (1975). Animal production in relation to conservation and recreation. *In* Proc. III World Conf. Anim. Prod. (Ed. R. L. Reid). Sydney University Press, Sydney.

Ulrich, G. F. and Parkes, A. S. (1978). The green sea turtle (*Chelonia mydas*): further observations on breeding in captivity. *J. Zool., Lond.* **185**, 237–251.

UNESCO/UNEP/FAO (1979). Tropical Grazing Land Ecosystems. UNESCO, Paris. pp. 655.

Walsingham, J. M. and Bather, D. M. (1976). Personal communication.

Wit, C. T. de (1975). Agriculture's uncertain claim on world energy resources. *Span* **18**, 2–4.

Young, N. E. and Newton, J. E. (1975). "Grasslambs", Grassland Research Institute Farmers Booklet No. 1.

Index